Highway Construction and Maintenance
Second Edition

Highway Construction and Maintenance

Second Edition

John Watson B.Sc., M.I.C.E.

Longman Scientific & Technical
Longman Group Limited
Longman House, Burnt Mill, Harlow
Essex CM20 2JE, England
and Associated Companies throughout the world

Copublished in the United States with
John Wiley & Sons, Inc., 605 Third Avenue, New York NY 10158

First published 1989
Second edition 1994

British Library Cataloguing in Publication Data
A catalogue entry for this title is available from the British Library

ISBN 0-582-23412-3

Library of Congress Cataloging-in-Publication data
A catalog entry for this title is available from the Library of Congress.

ISBN 0-470-23410-5 (USA only)

Set by 6 in 10/12 Times
Produced through Longman Malaysia, TCP

Contents

Acknowledgements

I am grateful to the following for permission to reproduce copyright material:

The British Cement Association for Table 3.2; Professor SF Brown for Fig. 2.7; Don & Low Ltd for Fig. 2.13; ELE International Ltd for Fig. 2.4; Halcrow Pavements Group for Fig. 9.5; the Controller of Her Majesty's Stationery Office for Figs 1.7, 2.3, 2.5, 4.3, 4.5(a), (b) and (c), 4.6, 5.1, 6.2, 8.3, 8.12, 9.3, 9.7, 9.8, 9.9 and Tables 8.1, 8.2 and 8.3; the Water Authorities Association for Table 7.4; the British Standards Institution for extracts from British Standards, complete copies obtainable from BSI, Linford Wood, Milton Keynes MK14 6LE; and other material as acknowledged in the text.

List of abbreviations

AADF	Annual Average Daily Flow – the total number of vehicles which use a road in one year, divided by 365
AAV	Aggregate Abrasion Value – an indication of an aggregate's ability to resist wear
CBM	Cement Bound Material – a mixture of cement and soil or aggregate, weaker than pavement quality concrete
CBR	California Bearing Ratio – a measure of a soil's ability to resist deformation caused by locally applied forces
CRC	Continuously Reinforced Concrete
CRCB	Continuously Reinforced Concrete Base
CRCP	Continuously Reinforced Concrete Pavement
C40, C20 etc.	An indication of the characteristic strength of a concrete mix, expressed as the 28-day characteristic cube strength, measured in N/mm^2
DBM	Dense Bitumen Macadam
DoT	Department of Transport – the UK government department responsible for public highways
d_{90}, d_{50} etc.	A measure of the soil particle sizes capable of being retained by a geotextile
E	The elastic modulus of a material – the ratio of stress to strain in that material
FWD	Falling Weight Deflectometer – a dynamic testing technique for pavement condition monitoring
ggbfs	Ground granulated blast-furnace slag
Hz	Hertz – cycles per second

i (suffix)	Initial characteristics – typically the penetration or softening point (P_i or SP_i) of a bituminous binder before it has undergone certain changes commonly experienced during laying and use
JCP	Jointed Concrete Pavement
MCV	Moisture Condition Value – a measure of the suitability of soils for use as engineering materials
mcv	Million commercial vehicles – a measure of the amount of damaging traffic using a pavement during a period
msa	Million standard axles – a measure of the damaging effect of traffic using a pavement during a period
OPC	Ordinary Portland Cement
O_{90}	A measure of the pore sizes in a geotextile
P, pen	Penetration – an indication of the viscosity of a residual bitumen or similar at a standard temperature, usually 25 °C
pfa	Pulverised fuel ash
pH	A measure of the acidity or alkalinity of a solution. A pH less than 7 indicates acidity while a value above 7 indicates alkalinity
PI	Penetration Index (of a bituminous binder) – an indication of the sensitivity of a bituminous binder's stiffness to changes in temperature
PI	Plasticity Index (of a soil) – the difference between the liquid and plastic limits, each expressed as a percentage
PQ	Pavement Quality – a general type of concrete
PSV	Polished Stone Value – an indication of an aggregate's ability to resist polishing and to retain surface roughness (microtexture)
r (suffix)	Residual characteristics – typically the penetration or softening point (P_r or SP_r) of a bituminous binder after it has undergone certain changes commonly experienced during laying and use
RB	Residual Bitumen – a bitumen obtained from the distillation of crude oil without further treatment
SEA	Sulphur Extended Asphalt – an asphalt in which sulphur replaces some of the bituminous binder

SP — Softening Point – an indication of the temperature at which a bituminous binder will have a certain arbitrary viscosity

STV — Standard Tar Viscosity – a measure of the viscosity of a liquid. The time in seconds taken for a standard quantity of the liquid to flow through a standard orifice in standard conditions

TRL — Transport Research Laboratory (formerly the Transport and Road Research Laboratory) – a major UK research centre

VDF — Vehicle Damage Factor

$$\text{VDF} = \frac{\text{Average damaging effect of each commercial vehicle}}{\text{Damaging effect of one standard axle}}$$

Commercial vehicle – any vehicle of weight greater than 1.5 tonne Standard axle – a notional two-wheeled axle carrying 8200 kg

VMA — Voids in Mixed Aggregate – the proportion which the volume of binder plus the volume of air voids bears to the volume of a sample of bituminous material as a whole, expressed as a percentage

1. Basics

1.1 Historical note

The first road builders of any significance in western Europe were the Romans, to whom the ability to move quickly from one part of the Empire to another was important for military and civil reasons. Roman roads are characterised by their linearity and, in popular perception, by their durability. A good alignment was sought since this provides the most direct route and since the risk of ambush in hostile territory is reduced. It was for this reason that the surface of the road was often elevated a metre or more above the local ground level – to provide a clear view of the surrounding country; hence the modern term 'highway'. The durability of such pavements is less absolute but nevertheless far exceeds anything achieved for many centuries after the fall of the Empire.

A typical major Roman road in the UK consisted of several layers of material, increasing in strength from the bottom layer, perhaps of rubble, through intermediate layers of lime-bound concrete to an upper layer of flags or stone slabs grouted in lime. The total thickness of such a pavement would be varied according to the ground conditions. In sound ground a thickness approaching one metre might be used; elsewhere this would be increased as necessary.

During the Dark Ages, and indeed well after that, no serious attempt was made in the UK to either maintain or replace the Roman road network, which consequently deteriorated. By the end of the Middle Ages there was in practice no road system in the country. Such routes as existed were unpaved tracks, swampy and impassable for most of the year and dusty and impassable for the remainder. Diversions around particularly poor lengths of road, private land or difficult topography had resulted in sinuous alignments. The general lawlessness combined with these characteristics to discourage all but the most determined travellers.

The first small change in this state of affairs was brought about by an Act of 1555 which imposed a duty on each parish to maintain its roads and to provide a Surveyor of Highways. As this post was unpaid and under resourced, and as the technical skills to match the task in hand did not exist, the obvious expectation that the post of Surveyor was unpopular and ineffective is generally correct.

This lack of resources remained a problem for over a century. In the latter part of the seventeenth century the first experimental lengths of turnpike road were established on the Great North Road (now the A1 trunk road). Turnpiking is a toll system whereby travellers pay for the use of the road. In the first part of the next century Parliament produced a series of Acts which enabled the establishment of Turnpike Trusts on major routes throughout the country. In this improved financial climate roadbuiling techniques gradually developed through the work of such pioneers as Metcalf, Telford and the eponymous Macadam. By about 1830 a system had evolved of well-paved roads of such quality that they imposed little or no constraint on road traffic. Journey times were limited not by the state of the road but by the nature of road vehicles.

The next improvement in the speed and cost of travel came about as a result of a radical change in vehicle technology – the building of the railways. The effect of this was to reduce road traffic between towns to such a low level that the turnpike system became uneconomic. Although road building in towns continued the Turnpike Trusts collapsed. Legislation in the late nineteenth century set the scene for the current administrative arrangements for highway construction and maintenance but the technology remained empirical and essentially primitive. Only in recent years has that situation changed to any great extent.

1.2 The aims of highway engineering

For economic activity to take place, people, goods and materials must move from place to place. The necessary movement has to some extent always been possible, but the growth in economic activity which characterised the Industrial Revolution in eighteenth-century England, and which has occurred or is occurring throughout the world since then, placed demands on the transport system which in its original primitive form it was quite unable to meet. This system developed to meet the new needs much more rapidly than it had previously, the economy expanded further, generating more traffic, and in this interactive way were produced canals and turnpike roads, then railways and most recently a network of modern roads.

Economic growth usually concentrates in areas where transport facilities are good – for example the construction in the UK of a motorway network during the quarter century starting in about 1960 has improved access from formerly remote areas to the capital and to international links, and those areas have prospered. In the previous century the railways had a similar effect; areas previously several days travel from any centres of population were, with the opening of a connecting railway, suddenly only a few hours away, and benefited as a result.

Roads provide a key element of the infrastructure whose function is to promote economic activity and improve the standard of living of the population. Highway engineering is concerned with the best use of resources to ensure that a suitable network is provided to satisfy this need of an economically sophisticated society.

1.3 Paying for roads

If a new road is to built, someone must pay for it. Generally, roads will be built either on behalf of a government agency as part of the public road network, or they will form part of a development project. Sometimes the interests of public service and development will coincide, in which case the cost may be shared by the government agency and the developer.

1.3.1 Public service roads

Public funds for roads to be built primarily as part of the public network are often prioritised on the basis of cost–benefit analysis. The costs of the road are often taken to include not only the construction cost of the road itself, but also the cost of measures necessary to mitigate its environmental side-effects. Land costs and professional fees are also included. Benefits include savings from improved journey times and from a reduction in the number of traffic accidents that would otherwise be expected during the life of the road.

Since these costs and benefits arise over many years they are often discounted to current equivalent value at the time of opening the road in order to simplify the decision making process. Competing schemes can then be compared on the basis of the relationship between their discounted cost and the discounted benefit which they are expected to provide. Broader economic benefits that the road might bring are often ignored.

Privately funded toll roads add to this process a commercial dimension in two ways. The project must bear any interest charges on outstanding loans borrowed to pay for the road, and the income which the road produces will of course depend on the amount of traffic using the road and the level of charges made.

1.3.2 Development roads

Here the road itself does not provide income directly, but it is part of the essential infrastructure needed to make a development marketable. The property development process is illustrated in its simplest form by Fig. 1.1.

Since this process takes place over a period of several years, funding of the roads must fit within the overall project requirements of adequate profit and acceptable cash flow. These are related through the cost of borrowing. Table 1.1 shows an example of a simple investment appraisal model applied to a development site. Note that the servicing cost includes the cost of the highway works and of providing utilities on the site. The cost of providing access might be for a road junction outside the site.

In the example the developer makes a profit of £118,000. This has been achieved by staging the construction works over the five-year period. If all the construction work had been carried out in the first year the interest charges arising from the high early cash flow would have made the project unprofitable.

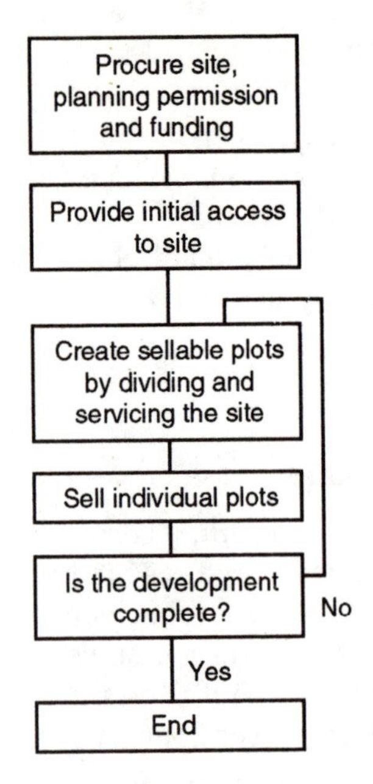

Figure 1.1 Outline of the property development process

Table 1.1 A simplified investment appraisal model for servicing a greenfield site

The site is to be bought, provided with roads and services and sold in small lots over a period for others to erect buildings and occupy individual plots. A new major road junction is needed to provide access to the site.

Gross area:	15 ha
Professional fees:	
Obtain planning permission	£10,000
Design and construction supervision	£5,000 per hectare
Marketing	£3,000 per hectare
Area of roads, verges and paths:	2 ha
Purchase cost of site:	£25,000 per hectare
Cost of access to site:	£250,000
Cost of roads and services within site:	£65,000 per serviced hectare
Market value of serviced land:	£150,000 per hectare
Borrowing interest rate	10%

In the table, values are in thousands of pounds

Costs	*Year 1*	*Year 2*	*Year 3*	*Year 4*	*Year 5*
Site purchase	375	0	0	0	0
Professional fees	34	24	24	34	8
Construction cost	445	195	195	195	65
Hectares serviced	3	3	3	3	1
Interest on previous year's balance	0	85	71	55	37
Costs for year:	854	304	290	274	110
Less sales revenue	0	450	450	450	600
Hectares sold	0	3	3	3	4
Profit (loss) for year	(854)	146	160	176	490
Cumulative balance	(854)	(708)	(548)	(372)	118

1.4 The nature of a pavement

The highway engineer is concerned with the provision of a safe, stable and durable surface over which traffic may move.

Originally roads were little more than tracks across the countryside and were hard, dry and dusty in summer and sodden and impassable in winter. The practice arose, initially in towns, of paving the surface of the road with resilient naturally occurring materials such as stone flags, and such a surface became known as a pavement. Today this term is applied to any surface intended for traffic and where the native soil has been protected from the harmful effects of that traffic by providing an overlay of imported or treated material. The purpose of providing this protection is to enable traffic to move more easily – and therefore more cheaply or quickly – along the road.

A modern pavement consists of a number of elements. These have various functions which contribute to the ability of the pavement to remain safe, stable and durable for a period of time and under the action of the weather and of large numbers of vehicles.

The surface of a pavement is of course that part which most immediately affects traffic. The needs of traffic are that the surface should be sufficiently uniform to allow traffic to pass in comfort and safety at reasonable speeds, that it should not be so slippery as to allow vehicles to skid in wet weather, and that it should be sufficiently free-draining to avoid intrusive spray or pools of standing water in wet weather.

Below the surface will lie one or more layers of imported or treated material whose function is to support the surface and to protect the soil beneath the pavement. These layers must be capable of withstanding the aggressive effects of traffic and of climate, and should protect the underlying soil from loads it cannot support. The designer should also ensure that the lower layers of the pavement are able to provide an adequate construction platform in order that the more highly stressed upper layers can be properly built and that the underlying soil is not disturbed unnecessarily during construction.

The picture is thus developed of a pavement which consists of several layers of material. In the most general case, pavements are made up of one or more layers of bound material – usually bound by either cement or a bitumen-based material – placed on one or more layers of an unbound granular material. An example is shown in Fig. 1.2 in which the bound layer consists of a concrete slab, the unbound layer is a subbase of graded stone, and the underlying soil is known as the subgrade.

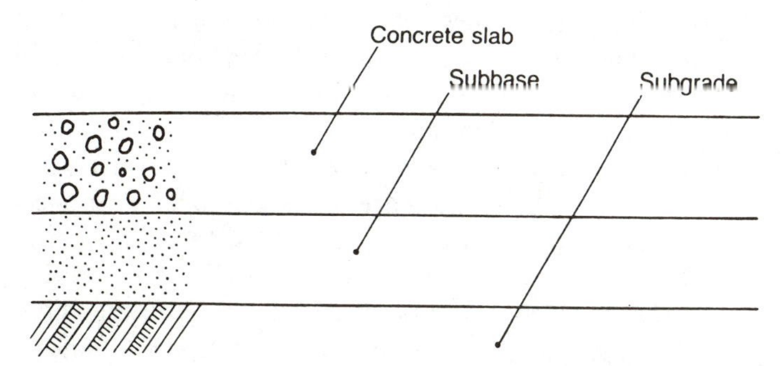

Figure 1.2 Section through a concrete pavement

This is an example of a rigid pavement – one in which there is no allowance for small deformations at the surface. The alternative to this is the flexible pavement, a general name applied to those pavements which are capable of retaining their structural integrity even when small vertical movements take place at the surface. Both types are suitable for most applications and the most lightly trafficked roads needs be no more complex in cross-section than that shown in Fig. 1.2. In contrast, flexible pavements built on poor soil and intended for heavy use may consist of a relatively large number of layers as shown in Fig. 1.3.

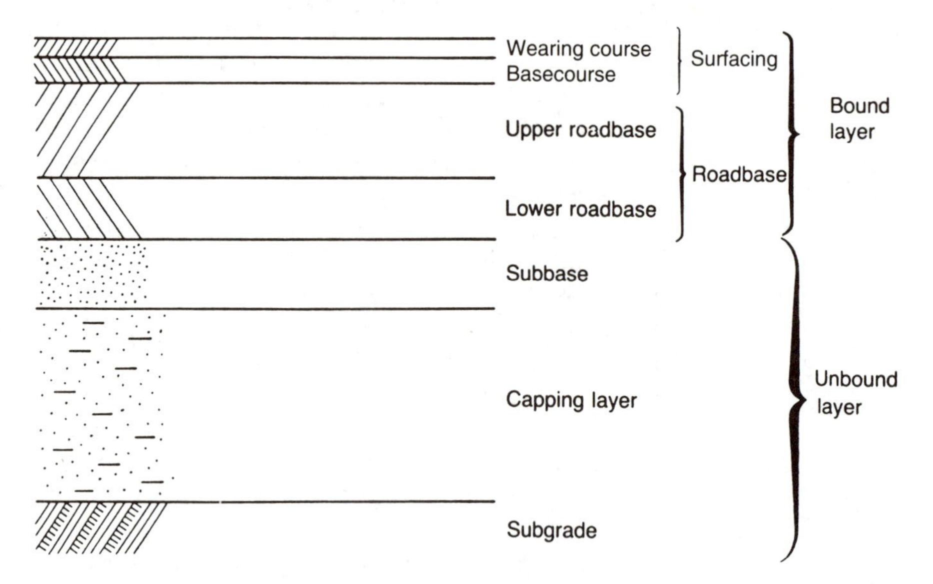

Figure 1.3 Section through a flexible pavement

The functions of the various layers in the pavement shown in Fig. 1.3 are as follows:

- *Capping and subbase* The pavement foundation – the purposes of these layers are to provide a working platform on which the roadbase may be built; to regulate the surface on which the road is to be formed; and to insulate the subgrade against the action of cold weather. There may also be a requirement for these layers to function as a temporary haul route during the construction of subsequent lengths of capping and subbase, and during roadbase and surfacing operations. The design of these parts of the pavement is discussed in Chapter 3.
- *Roadbase* The roadbase makes the major contribution to the strength of the pavement. Its functions are to distribute loads applied at the surface so that excessive stresses are not transmitted through the pavement foundation to the subgrade, and to provide a sufficiently stable base to support the surfacing. Depending on circumstances and the choice of the designer, the roadbase may consist of cement- or bitumen-bound material, stabilised soil or carefully graded granular material.

- *Surfacing* This layer should provide a suitable surface for vehicular traffic and, in most cases, prevent the ingress of surface water into the pavement. The needs of traffic will vary from site to site. Requirements which are often considered include surface regularity, surface texture, durability and flexibility. Surfacing is often provided in two layers, known respectively as the base course and the wearing course. The base course is essentially an extension of the roadbase but by virtue of the nature of the material can be laid to finer tolerances than can most roadbase materials. The wearing course further regulates the surface and in addition should provide the general properties of the surfacing described above. Roadbase and surfacing design is discussed in Chapter 4.

In the case of the concrete pavement illustrated in Fig. 1.2, the functions of the roadbase and surfacing are combined in the concrete slab. Note however that the way in which the two pavement types discharge these functions are very different. Concrete pavements are discussed in Chapter 5.

The designer of a new pavement must ensure that each layer is capable of satisfying these requirements. In so doing, decisions must be made about the nature of the materials to be used, and the thickness of each layer. These decisions lie at the heart of the highway engineering.

If we are to do this, it is helpful to understand the consequences of incorrect decisions. We need to know how a pavement is likely to fail.

1.5 Modes of failure

Taking a very broad view, there are two types of pavement failure.

Failure of the surfacing may take the form of loss of surface texture, loss of surface regularity or loss of impermeability. In the knowledge that such failure is not deep seated, remedial measures will consist of repairs to or replacement of the surfacing only. Such failures can often be prevented or greatly postponed by a careful choice of surfacing materials.

It is rather more difficult to diagnose and to remedy failure below the surface. Figure 1.4 illustrates a flexible pavement whose structure has been simplified to one bound and one unbound layer and which is subjected to a vertical wheel load.

Modes of failure illustrated in Fig. 1.4 are (A) horizontal strain at the bottom of the bound layer, causing the upward propagation of vertical cracks through this layer; and (B) vertical strain at the top of the subgrade, causing consequent deformation in the bound and unbound layers and possibly leading to deformation at the surface. Analogous failure modes may arise with concrete pavements, with the exception that vertical subgrade strain will not be a primary cause of surface deformation, although it may in extreme cases lead to undermining and cracking in the concrete slab.

Tensile failure at the bottom of the bound layer may be avoided either by the provision of sufficient thickness of bound material to contain horizontal stresses within acceptable limits for a given material, or by providing a material of sufficient strength to resist the stresses which will arise in a bound layer of

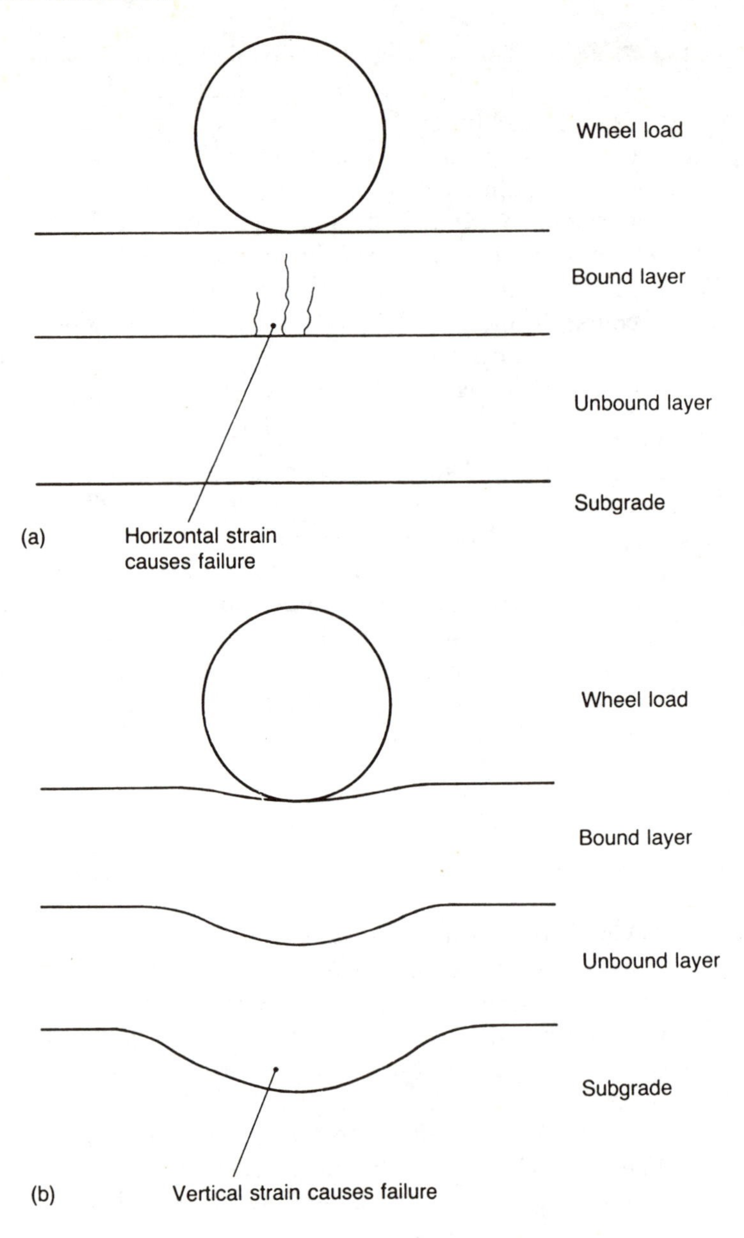

Figure 1.4 Examples of failure modes

sufficient thickness. In practice the designer seeks to achieve a balance between these two approaches.

Where the unbound layer is of insufficient depth or is of deficient material, or where the bound layer fails to distribute adequately the loads imposed on it, permanent vertical strain may occur at the top of the subgrade (that is, at the formation). Furthermore, the use of the unbound layer by construction or other traffic and its ability to withstand this must be considered. This is discussed further in Chapter 3.

1.6 Causes of failure

Only in the most extraordinary circumstances is it the case that a new pavement is caused to fail by the application of a single wheel load, of whatever magnitude. The general experience is that repeated application of loads results in the repeated development of stresses tending to cause failure, and that the ability to withstand strain progressively diminishes with the increasing number of load applications. When failure occurs, it is as a result of material fatigue. This concept is the basis of current design practice.

If failure occurs in this way, a rational design approach must be based upon a study of the responses of the pavement to repeated cyclical loading, on the nature of that loading, and on the number of load applications which can take place before the pavement fails. This introduces the notion of a pavement having a predetermined and finite life. In order to control costs the designer may decide for how long a pavement should last. To ensure a consistent approach we should decide at what stage in its deterioration a pavement is deemed to have failed.

1.7 Pavement life

In a few cases the required life of a pavement is determined by external circumstances; for example, a temporary road to serve a major construction site might be designed to last for the duration of the construction works and no more. Generally however the matter is less clear-cut.

The life of a pavement may be extended almost indefinitely by judiciously applied maintenance works – particularly by the provision of overlays. Techniques are available (Chapter 9) by which the condition of a pavement may be assessed and from which deductions may be made regarding the timing and nature of necessary repairs. An aim of the maintenance engineer is to identify the onset of critical conditions in the pavement – those beyond which rapid structural deterioration is likely, and major works are necessary to restore the road to a sound condition and increase pavement life. Once this critical stage has been passed an overlay can no longer be relied upon to provide a substantial increase in pavement life; expensive reconstruction is likely to become necessary.

A survey has been carried out of the correlation between the structural condition of a flexible pavement and defects apparent at the surface[1] (see also Table 9.3). This survey indicated that the critical stage described above is generally consistent with 10 mm rutting in the wheel paths or the beginning of cracking in the wheel paths. Current UK practice is to provide an overlay when these conditions, or others consistent with them, arise.

In designing a new pavement we therefore define the design life as being that period of time likely to elapse before this critical condition will arise. Current UK design methods are based on this definition.

If such a definition of design life is to be of practical value the designer should be aware of the implications of different choices of design life. It is possible to design a pavement with a design life of say five, 40 or any other number of years but, in the situation where the need for a pavement is expected to remain for ever, there are arguments for and against any particular choice of pavement life.

For example, if we design a pavement to last for five years before the onset of critical conditions, the pavement may reasonably be expected to be thinner and cheaper than one planned to last for 40 years. At first sight then the cost of providing a short-life pavement may be expected to be less than that of providing a long-life pavement. When the cost of maintenance works is included in the reckoning a rather less clear-cut picture emerges.

Works which might be expected in ideal circumstances during the first 40 years of the life of a road paved in flexible materials are suggested in Table 1.2. This table is based on the approximate design assumption that reconstruction will become necessary when the design life has been exceeded by about 50 per cent. The costs which the highway authority will have to bear include the initial construction and works costs of all maintenance, together with the cost of traffic control during maintenance operations. Additionally, a cost will accrue to the road users in the form of delays during roadworks.

Cost comparisons are further complicated by the way in which the expenditure varies throughout the 40 year period, depending on the design life of the pavement. This is a common problem in the costing of alternative business strategies, and a standard accounting procedure known as the calculation of discounted costs exists whereby the cost of activities planned for different times in the future may be compared by discounting to an equivalent present cost (see any standard accounting work or reference 2).

Table 1.2 Idealised programmes of works to pavements of various design lives

	Design life of pavement (years)			
Year	*5*	*10*	*20*	*40*
0		CONSTRUCTION		
5	*	–	–	–
10	Reconstruct	Resurface	*	*
15	*	*	*	*
20	Reconstruct	Reconstruct	Resurface	*
25	*	–	*	*
30	Reconstruct	Resurface	*	*
35	*	*	*	*
40	Reconstruct	Reconstruct	Reconstruct	Resurface

* Minor surface maintenance, such as patch, surface dressing.

Taking all these factors into account, it is possible within reasonable limits of accuracy to compare the costs of providing otherwise similar pavements of different design lives. Results of such an analysis are shown in Fig. 1.5. Variables which affect the form of the curve shown include the amount of traffic using the road, variations in this quantity during the period considered, and the discount rate used in calculating equivalent present costs; however, in practice

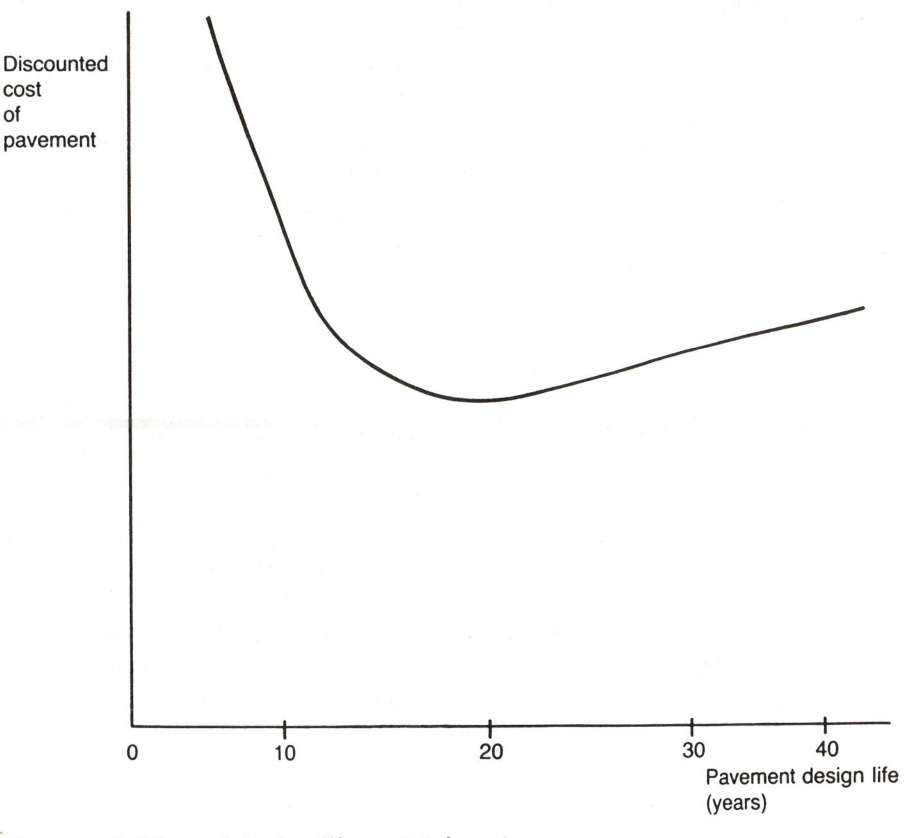

Figure 1.5 Effect of design life on total cost

none of these is likely to alter significantly the conclusion to which Fig. 1.5 leads us. Flexible pavement designed to last 20 years before the onset of critical conditions will give substantial overall cost savings in comparison with those designed to last for shorter periods, and marginal cost savings in comparison with those designed to last for longer. Permanent flexible pavements are generally designed for a life of 20 years.

Similar studies of the performance of rigid pavements point to the use of a 40 year design life in these cases.

1.8 The influence of traffic

We have seen that the failure of a pavement may be caused not merely by the passage of time but rather by the number and size of the loads applied to it. All other things being equal, we may therefore expect a pavement trafficked by large numbers of heavy vehicles to reach critical conditions before one which is less heavily trafficked. Before we can proceed with the design of a pavement we should therefore know not only how long we want the pavement to last, expressed as so many years before the onset of critical conditions, but also the volume and nature of the traffic that will use the pavement during that period.

1.8.1 Estimation of wheel loads

During its life a pavement will be subjected to wheel loads of widely ranging magnitudes. Some of the most severe loads will occur during construction, especially during compaction by steel tyred rollers which produce very high stresses. Loads imposed by normal traffic with pneumatic tyres will be much less concentrated, the contact pressure between wheel and road being equal to the pressure to which the tyre is inflated. Nevertheless, such loads may often be of similar magnitude to those imposed during construction, and will probably be much more numerous. The applied loads will range from a matter of kilograms, in the case of a bicycle, to several tonnes, in the case of a large commercial vehicle. The enthusiastic designer might seek to allow for all of these in a detailed appraisal of traffic behaviour during the life of the pavement. Such enthusiasm would be misplaced.

Experience has shown that the damaging effect of a loaded wheel passing over a pavement varies in approximate proportion to the fourth power of the load carried by the wheel. If we were to adopt as our unit of damaging effect, the damaging effect of the smallest likely wheel load (that of a 20 kg bicycle ridden by a person weighing say 60 kg) then a comparison of the damaging effects of differently loaded wheels might be as shown in Table 1.3. Allowing for the number of wheels per vehicle, each hypothetical four wheeled, sixteen tonne lorry that passed over a pavement would cause the same amount of damage to the pavement as would rather more than sixty-five thousand cars, or two hundred million bicycles.

Table 1.3 Relative damaging effects of various vehicles

Vehicle	*Wheel load*	*Damaging effect of one wheel load (relative to one bicycle wheel)*
Bicycle	40 kg	1
Car	250 kg	1525
Lorry	4000 kg	100,000,000

Since public roads may be expected to carry large lorries and other heavy vehicles it is clearly in that class of traffic that the major cause of pavement damage is to be found. For this reason, and for reasons of mathematical convenience, the standard unit of damaging effect is related not to the lightest vehicle to use the road, but to an arbitrary load of the same order of magnitude as the lorry considered in Table 1.3. The standard unit of the damaging effect of a vehicle on a pavement is known as the standard axle, and is such that a standard axle is one which imposes a load of 8200 kg (about 80 kN) on the pavement by means of two parallel 40 kN wheel loads. Because of their relative insignificance, loads imposed on the surface by vehicles other than commercial and other heavy vehicles are ignored. Only vehicles of total weight greater than 1.5 tonnes are considered, and for the sake of convenience all such vehicles are known as commercial vehicles.

It then remains for the amount of damage caused to the pavement by each commercial vehicle to be related to that caused by the standard axle. This ratio is known as the vehicle damage factor (VDF) and is such that

$$\mathrm{VDF} = \frac{\text{Damaging effect of each commercial vehicle}}{\text{Damaging effect of one standard axle}}$$

For any particular vehicle of known axle loads it is a simple matter to calculate the VDF; one simply calculates the damage factor for each axle, using the fourth power relationship mentioned above, and sums these to obtain the VDF. An example is given in Fig. 1.6.

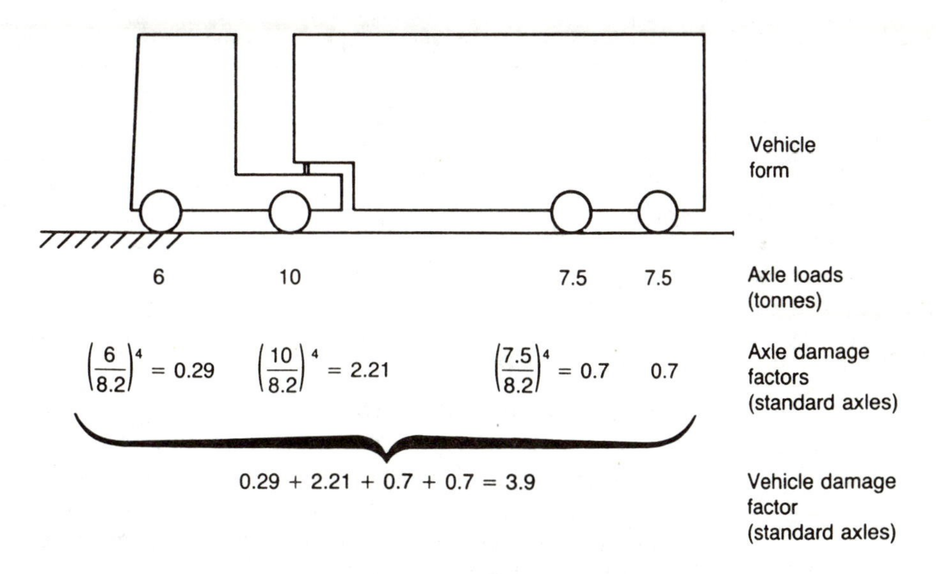

Figure 1.6 Example of calculation of vehicle damage factor

It is clearly unrealistic to attempt to carry out such calculations for all vehicles in all conditions of loading which are expected to use a public road during the next 20 years. A more general approach is required.

A long term survey of the axle loads of commercial vehicles on public roads in the UK[3] has found the average VDF of commerical vehicles to vary with time, increasing from year to year; and to vary with the total volume of commercial vehicles using the road in question – roads used by large numbers of commercial vehicles exhibiting a high average VDF. The results of this survey have been analysed[4] to produce the following relationship:

$$\mathrm{VDF} = \frac{0.35}{0.93^t + 0.082} - \frac{0.26}{0.92^t + 0.082} \cdot \frac{1.0}{3.9^{F/1550}} \qquad [1.1]$$

in which t = number of years from the base year, 1945, to the year in question (thus for 1995, $t = 50$)

F = 24 hour AADF (see below) of commercial vehicles during the year in question

Vehicle damage factors calculated from this equation are indicated in Table 1.4. These factors are applied to all commercial vehicles in excess of 1.5 tonnes which are expected to use the new pavement. It is often considered to be sufficiently accurate to obtain the VDF appropriate to the year in which the pavement achieves half its design life (year 10 in the case of a pavement with a design life of 20 years), and to apply this factor globally in converting predicted numbers of commercial vehicles to preduced numbers of standard axles.

Table 1.4 Vehicle damage factors

	Daily flow of commercial vehicles AADF			
Year	*250*	*1000*	*2000*	*4000*
1995	1.18	2.12	2.76	3.14
2000	1.22	2.31	3.00	3.40
2005	1.34	2.47	3.18	3.60
2010	1.43	2.60	3.33	3.76
2015	1.51	2.69	3.44	3.88

1.8.2 Estimation of traffic volumes

Having estimated the average damaging effect of each commercial vehicle expected to use the new pavement, we may prepare an adequate estimate of the number of such vehicles likely to pass over any particular point in the pavement during its design life. We are therefore concerned to identify the total number of commercial vehicles moving in each direction along a two-way road; the total flow does not concern us.

Generally there are two stages of this process. The first is the estimation of the expected rate of passage of such traffic deduced directly from current information. Typically, one of three circumstances will obtain.

In cases such as the reconstruction of a road where there is no proposed change to the form of the road network and where no changes are expected in the nature or volume of the traffic currently using the road, traffic flows measured at the site may be used as the basis for forward estimation. Attention should be paid to the nature of the survey information from which any conclusions are drawn.

The designer is concerned with the total number of commerical vehicles likely to use the road over a long period. It is unreasonable to expect traffic flows to remain constant in time at any particular location; rather, one may expect flows to vary according to the time of day, the day of the week, and the week of the year. The nature of this variation will depend on the situation of the road. Traffic flows on a main suburban commuter route may vary throughout the day very differently from those on the access to a large shopping area; traffic volumes in a large inland city may fall during the summer holidays and rise during the Christmas shopping period while the converse may be true in a seaside resort. To present as true a picture as possible of long term traffic conditions the practice is to take as the basis of design the Annual Average Daily Flow (AADF) of commercial vehicles, that is to say the total number of commercial vehicles which use the road per year, divided by 365. Sometimes such information will be available to the designer but more often arrangements will have to be made for a suitable traffic survey to be held. Such a survey should be continuous over several days, should be taken at a time of the year when flows are 'typical' for the area – often during the spring or late autumn – and should of course distinguish commerical vehicles from other types of traffic. Automatic traffic counters are available which by means of inductive loops set in the surface of the road, produce hourly counts, classified by length, for periods of several weeks at a time. Vehicles in excess of 5.5 metres may be considered to be commercial vehicles for pavement design purposes. Note that the requirements of pavement design in respect of traffic data differ from those of traffic engineering where the peak hours are often of more concern.

In many cases however the provision of a new pavement will be associated with changes in the form or capacity of the road system. For example, a completely new route may be provided, or the capacity or length of an existing route may be altered. Experience has repeatedly shown that where the attractiveness of a route – often measured in terms of journey time – is improved, the amount of traffic using it increases; and that this increase is due not only to the transfer of traffic from other parallel routes, but also to new traffic using the route to make journeys that would not have been considered worthwhile before the change was made. This is a natural consequence of the role of roads in promoting economic development. It is possible to develop mathematical models based on information obtained from surveys of the origins and destinations of traffic passing through the area in which the route lies, and various computerised methods of analysing such a situation exist, but the details of such methods are complex and lie beyond the scope of this book.

Land use changes may also affect traffic flows – for example, where an area of previously agricultural land is developed for industrial use. This is an area which has been frequently studied and adequate means are available for such local traffic effects to be predicted for pavement design purposes.[5,6]

By one or a combination of these means an estimate may be made of commercial vehicle flows during the year to which the collected or predicted data relate – the base year. It then remains to calculate from this base year figure the total number of commercial vehicles (and hence the number of standard axles) which will pass over a point in the pavement during its design life. If no allowance is to be made for future changes in traffic then this process reduces to simple multiplication, thus:

$$\text{Cumulative traffic} = \text{AADF(base year)} \times 365 \times \text{Design life (years)}$$

However, such an equation is often misleading because it makes no allowance for long-term variations in the amount of traffic. One can imagine circumstances in which no change in traffic is possible as the years go by, but generally changes in vehicle ownership, population or economic activity have resulted in changes in the usage of the road network. For example in the UK long-term traffic counts at various sites indicated an annual growth of 4 per cent per annum during the period 1982–1992. If we incorporate this notion of compound annual growth the situation becomes slightly more complex mathematically:

$$\text{Cumulative traffic} = \text{AADF(base year)} \times 365 \times \frac{(1+r)^n - 1}{r} \qquad [1.2]$$

in which n is the design life (years)

r is the annual rate of change in the AADF. (Note that r can be positive or negative.)

and base year flows relate to those in the first year of the road's life.

Equation [1.2] may be reduced to

$$\text{Cumulative traffic} = \text{AADF} \times K \qquad [1.3]$$

Values of K are given in Table 1.5.

Table 1.5 Values of *K* in equation [1.3]

Annual rate of change in AADF (r)	*Design life (years) (n)*				
	5	*10*	*15*	*20*	*40*
5%	2017	4591	7876	12069	44092
4%	1977	4382	7309	10869	34684
3%	1938	4184	6789	9808	27521
2%	1899	3997	6312	8869	22047
1%	1862	3819	5875	8037	17843
0	1825	3650	5475	7300	14600
−1%	1789	3490	5108	6646	12083
−2%	1753	3338	4771	6066	10116
−3%	1719	3228	4462	5551	8569
−4%	1685	3058	4178	5092	7342
−5%	1651	2929	3918	4683	6362

In cases where the AADF per traffic lane can be reliably estimated for the base year and where there is likely to be little transfer to other traffic lanes in the future – for example in the case of a single two-lane, two-way carriageway – this approach will give adequate results. However, where the possibility exists of significant volumes of traffic transferring to adjacent traffic lanes, as is the case where more than one lane is provided for each direction of traffic, then the proportion of the total commercial vehicle flow passing over a point in the nearside lane is found to fall as the total volume of traffic increases.[7] While at low flows effectively all commercial vehicles may be expected to remain in the nearside lane, as flows on a road where lane transfer is possible approach the practical capacity of the road so up to about 45 per cent of all commercial vehicles transfer to the lane adjacent to the nearside lane. The situation is complex mathematically and is most conveniently resolved by the nomogram in Fig. 1.7. As would be expected, Fig. 1.7 gives lower results than does equation [1.3] when modelling high flow rates; at low flows there is no significant difference between the two (see Fig. 1.8).

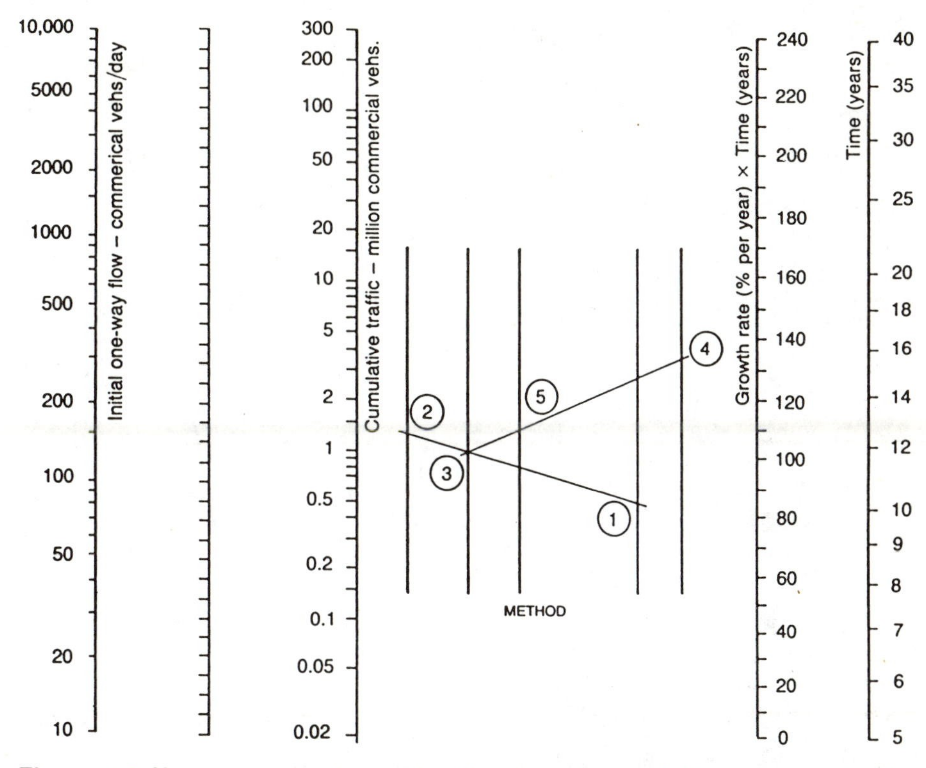

Figure 1.7 Nomogram for forward estimation of cumulative one-way traffic, nearside lane

In cases where there are large differences between cumulative flows in the nearside lane and those in the adjacent lanes, the opportunity is sometimes taken to reduce the thickness – and therefore the cost – of the middle lanes. If we can rely on equation [1.3] to predict the total number of commercial vehicles using all available traffic lanes, and if Fig. 1.7 reliably predicts the total number of commercial vehicles using the nearside lane, then the difference between the two predictions will be indicative of the total number of commercial vehicles expected to use the lane adjacent to the nearside. A design for the lanes adjacent to the nearside may be prepared on this basis.

1.8.3 Summary of prediction of the influence of traffic

To summarise, the process of predicting the damaging effect due to traffic that the pavement will be required to withstand consists of:

(1) Predicting the volume of traffic that would use the pavement if it were available for use in a base year, volume being expressed in terms of Annual Average Daily Flow of commercial vehicles;

(2) By means of equation [1.2] or [1.3] or by means of Fig. 1.7, according to the situation, estimating the total number of commercial vehicles that will pass over any point in the pavement during its design life;

(3) By means of equation [1.1], estimating the vehicle damage factor to enable the total equivalent number of standard axles to be predicted.

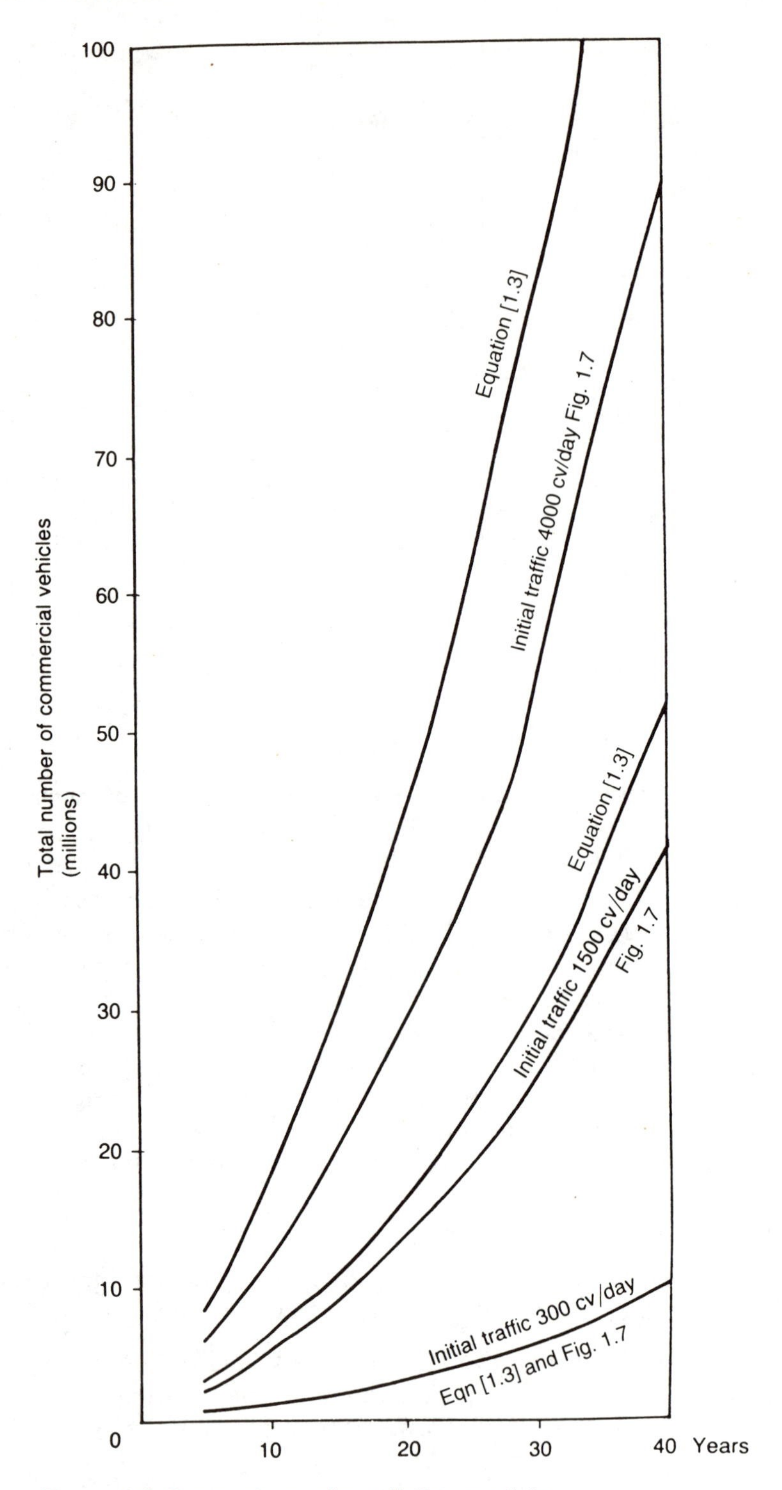

Figure 1.8 Comparison of predictive models

Pavements are designed to resist the damaging effect of this number of standard axles.

It is perhaps unfortunate that some statements of design practice[8] omit from the design process the work described in this chapter. The designer is instead invited to use standard design charts deduced from the work outlined above and an assumed traffic growth rate of 2 per cent per year. This may not always yield satisfactory results.

1.9 Elements of highway engineering

Figure 1.9 illustrates the main components of a highway and the considerations which lie behind it, relating these to the chapters of this book.

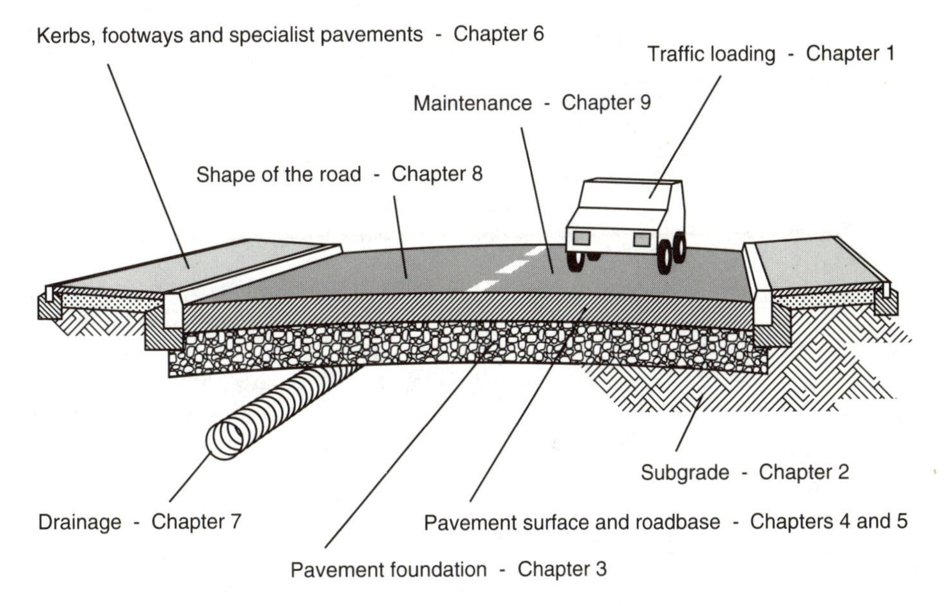

Figure 1.9 Elements of highway engineering

Revision questions

1 Consider the site development to which Table 1.1 refers.
(**a**) How could the development be made more profitable?
(**b**) Why is the total area of land sold less than the area bought?
(**c**) What benefits come from this project?

2 Explain why the nomogram used for the forward estimation of cumulative traffic is thought to give satisfactory results in most cases. Give an example of a situation in which it would be wrong to use the nomogram for this purpose. In your example state your preferred method for cumulative traffic prediction and explain why it is to be preferred.

3 Axle loads and payloads for two typical vehicles are shown in Fig. 1.10. Assess the relative damaging effects on the pavement of the 38 tonne and 32 tonne vehicles shown, per tonne of goods carried. Is a simple arithmetical calculation such as this sufficient to estimate the effect on road pavements of changes in the maximum allowable size of such vehicles? Why?

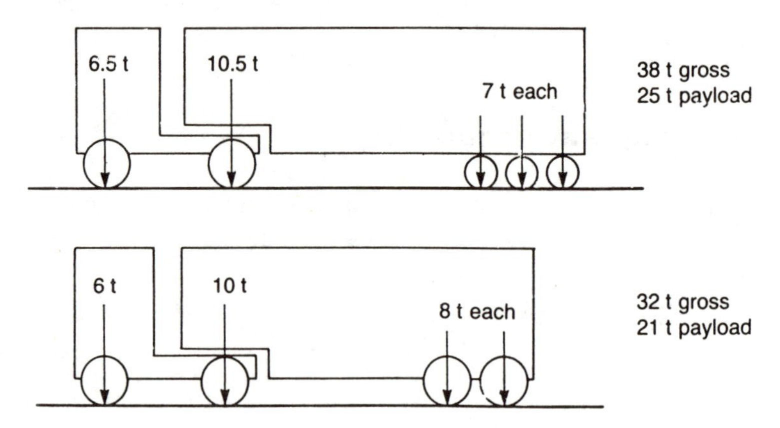

Figure 1.10 Two typical goods vehicles

4 Explain what is meant by a 'Standard Axle' and define the term 'Vehicle Damage Factor' (VDF) in the context of pavement design. A regression model has been developed for predicing VDF in the UK using as variables time, design life and commercial traffic volume. Why are these variables significant? Explain the limitations of this model in the UK giving one circumstance in which the model is inadequate. For this one circumstance, discuss how you might determine a realistic value for VDF.

References and further reading

1 Lister, N. W., 'Deflection criteria for flexible pavements', DoE/DTp, Transport Research Laboratory, Laboratory Report 375, 1972.
2 'An introduction to engineering economics', ICE.
3 Currer, E. W. H. and O'Connor, M. G. D., 'Commercial traffic: its estimated damaging effect 1945–2005', Transport Research Laboratory, Laboratory Report 910, 1979.
4 Addis, R. R. and Robinson, R. G., 'Estimation of standard axles for highway maintenance' Proceedings of Symposium of Highway Maintenance and Data Collection, University of Nottingham, 1983.
5 Bartlett, R. S. and Newton, W. H. 'Goods vehicle trip generation and attraction by industrial and commercial premises', Transport Research Laboratory, Laboratory Report 1059, 1982.
6 'Traffic generation: user's guide and review of studies', Reviews and research series No 25, Greater London Council, 1985.
7 Thrower, E. N. and Castledine, L. W. E., 'The design of new road pavements and of overlays: estimation of commercial traffic flows', Transport Research Laboratory, Laboratory Report 844, 1978.
8 Department of Transport: Departmental Standard HD 14/87 'Structural Design of New Road Pavments', Department of Transport, 1987.

2. Highway construction materials

2.1 Introduction

In the context of highway construction there are three general types of material with which the engineer is concerned.

The first group of materials is that which includes all the various types of soil which may lie below the formation. These naturally occurring soils can have widely varying properties. Different types of subgrade material will lead to different pavement designs, particularly in the lower layers. Different construction techniques may also be necessary, and the scope and cost of the construction project can be affected.

Secondly we will consider the unbound materials commonly used in the lower courses of the pavement. These materials may vary from naturally occurring ('as-dug') sands and gravels to graded synthetic aggregates such as clinker and slag.

The third and most varied group of materials consists of soils and graded stone which have been stabilised to a greater or lesser extent by the addition of a binder or other agent so as to substantially modify their properties. Such materials include, for example, stabilised soils and concrete.

2.2 *In situ* soils

Start digging with hand tools at almost any undeveloped site and the first layer of material exposed will be organic in nature – it will contain living organisms and their decomposing and decomposed arisings and will be mechanically weak. This topsoil is unsuitable for use for engineering purposes and is therefore set aside during construction, either for disposal or for re-use in planted areas upon completion of construction works.

Below the topsoil will lie an inorganic material whose properties may be reliably quantified and which is often suitable as a foundation for the pavement. This will sometimes be in the form of 'rock' – such as limestone, granite, sandstone and the like – but is more often found to be a soil. A common definition of 'rock' in this context is that the term shall include any hard natural or artificial material requiring the use of blasting or pneumatic tools for its removal but excluding small individual masses; 'soil' is, by default, any other material found below ground level. Note that these descriptions differ from those used by the geologist, to whom all naturally occurring non-fluid materials are rock.

A further sub-division is often made in the case of soils – material may be either 'suitable', that is to say it may be used as an engineering material somewhere in the works, perhaps as fill in an embankment, or it may be 'unsuitable' in which case it is considered to be insufficiently stable for re-use elsewhere except for the forming of non-loadbearing mounds. A typical application for unsuitable material is in the forming of mounds to act as visual and acoustic screening to a new road.

A naturally occurring soil will consist of particles of solid material, the voids between which are to a greater or lesser extent occupied by water. The physical properties of the soil will depend on the nature of these particles, the proportion that the volume of voids present bears to that of the whole body of material, and the amount of water present. In considering the suitability of a soil to support a pavement or other structure we are particularly concerned with the soil's ability to resist deformation caused by applied loads. Deformation may be caused by the soil tending to flow under the action of the applied load, or by its changing in volume.

Volume changes in soils resulting in settlement at the surface are caused by a limited rearrangement of the soil particles, resulting in a reduction in the proportion of voids present in the soil. Where the water or air originally present in the voids is able to leave the system quickly, as for example in the case of a coarse-grained soil with large voids between the particles, then such volume changes are not a great problem, since full settlement can be achieved by readily available techniques within a short period. In other cases, where the soil does not drain so freely, the water in particular can only be expelled from the system by the continuous application of a substantial load over a long period. The first form of settlement, which may be induced by the use of compaction plant, is known as immediate settlement; the second, consolidation settlement.

Immediate settlement is a property of coarse-grained soils such as sands and gravels, while consolidation settlement is characteristic of soils consisting of very small individual particles such as clay and silt. For this and other reasons the engineer is often concerned with the range of particle sizes present in a soil.

The particle size distribution of a soil is often determined by sieve analysis. A full description of such a test is given in BS 1377 'Methods of test for soils for civil engineering purposes'. The test apparatus consists of a series of sieves of gradually reducing standard sizes arranged vertically above one another with the coarsest at the top, all capable of being agitated by some mechanical means. The soil to be tested is dried and introduced to the top sieve, the whole stack of sieves shaken for some time, and the proportion retained on each sieve determined by subtractive weighing and expressed as a proportion of the mass of the whole. The results are reported by means of a table, or a cumulative frequency curve as in Fig. 2.1.

A knowledge of particle size distribution is of considerable value in assessing

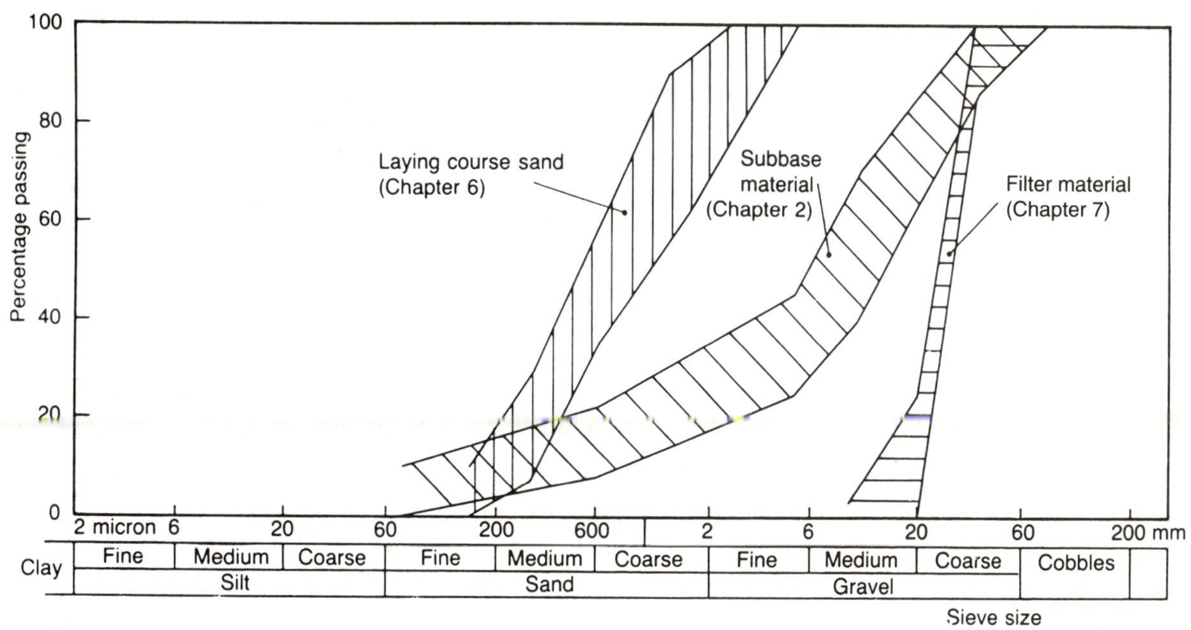

Figure 2.1 Typical grading requirements

the likely behaviour of the soil under a variety of circumstances. Soils with relatively large particles present free drainage paths for water in the soil and are therefore capable of being easily compacted; water can leave the system early. Very fine textured soils such as clays do not share this property and so are much less suitable for use as fill material, particularly in cases where early stability is important such as beneath pavements and other structures.

2.2.1 Granular soils

Free draining soils which consist of large numbers of discrete particles individually visible to the naked eye do not bind together in their natural state; the only internal forces which tend to resist the deformation of a mass of such soil by an applied external force are the frictional forces which exist between one particle and another. These soils are said to be granular and include all soils more coarsely grained than silt. If we are to select such a material as fill or as part of the structure of a pavement, as is often the case, we will therefore be concerned to ensure that these frictional forces are present to such an extent as to provide adequate stability. It is often necessary to pay some attention to the grading of the material, beyond the mere consideration that the particles should exceed a certain size; and to the shape of those individual particles.

An extreme example of a granular material is offered by a mass of uniform particles of equal size, such as a pebble beach. The characteristics of the surface of such a material are that it is very open in texture, easily disturbed by external forces, and very difficult to compact fully. Such a material is unsuitable as a platform upon which the construction of subsequent layers may take place because of both its lack of stability and its very open texture into which fines from superimposed layers may be lost. To effect an improvement one may seek to fill the voids between the coarse particles in the original uniform granular soil; a rather finer material is clearly required. If this finer material is itself single sized then the drawbacks originally encountered may again arise unless a further quantity of even finer material is included in the mix. In practice, this need results in the desirability of a continuously graded granular material for use as a high-grade fill beneath and within the pavement. Examples of specification requirements for sieve analysis results obtained with samples of various materials are shown in Fig. 2.1.

The shape of the particles can also be significant. Continuously graded naturally occurring aggregates are often available, and can be cheap and effective in some circumstances. However, these suffer from the disadvantage that they were formed by the rolling and abrading action of moving water which tends to result in more or less round shaped particles, with smooth surfaces. The individual particles thus slip relatively easily against one another, and a high degree of interlock is often difficult to obtain with these naturally occurring materials.

As an alternative to these natural aggregates, rock is often broken down to a suitable size and sorted to produce a material of suitable grading to enable a dense, stable mass to be achieved in the field. Crushed rock particles are by their nature angular in form and rough in texture; they therefore tend to exhibit a much higher degree of interlock than do otherwise comparable naturally occurring materials. Crushed rock is preferred to natural sands and gravels especially for use in the foundation of heavily trafficked pavements or where the requirements of construction traffic are likely to be partlcularly taxing. In addition to crushed rock, similar properties are obtained from crushed concrete, crushed blast-furnace slag or well-burnt non-plastic shale. In the UK such materials, with grading as shown in Fig. 2.1, are referred to as Type 1 granular material, in accordance with the Department of Transport Specification for Highway Works; natural sands and gravels graded to a less demanding specification, are referred to as Type 2 granular material.

Crushed rock and crushed slag also find use as a higher grade fill material suitable for use in the upper layers of lightly trafficked pavements provided that the more stringent grading requirements indicated in Table 2.1 are met, and provided that the moisture content of the material is properly controlled during its laying (see below). This material is known as wet-mix macadam – named after the famous engineer, and to distinguish it from dry-bound macadam.

The construction process by which dry-bound macadam is formed is such as to attempt to achieve the objective of filling with fines the voids between the coarse particles not by the provision of a premixed aggregate of an appropriate grading but rather by aiming to achieve the appropriate mix proportions *in situ*. Briefly, dry-bound macadam is formed by spreading coarse aggregate of nominal size 50 mm over the area to be treated in layers about 100 mm thick, and spreading over each layer of this coarse, open material a layer of fines graded from 5 mm down to dust. The whole is then agitated by vibratory compaction plant and the fines enter the voids within the coarse aggregate. Application of fines is repeated as necessary to refusal, the end

Table 2.1 Grading requirements of some granular materials[1]

	Percentages by mass passing for:		
BS Sieve size	*Subbase Type 1*	*Subbase Type 2*	*Wet mix*
75 mm	100	100	100
50 mm	–	–	100
37.5 mm	85–100	85–100	95–100
20 mm	–	–	60–80
10 mm	40–70	45–100	40–60
5 mm	25–45	25–85	25–40
2.36 mm	–	–	15–30
600 μm	8–22	8–45	8–22
75 μm	0–10	0–10	0–80

result being a stable, suitable mass of well compacted material. In order that the fines may easily penetrate the coarse aggregate it is important that they are kept dry.

2.2.2 Cohesive soils

Very fine grained soils exhibit the property that when slightly moist and squeezed in the hand they may readily be formed into coherent lumps; the soil particles tend to stick together. This property is not shared by granular soils. Soils which stick together in this way are said to be cohesive; the most common cohesive soil is clay.

In their natural state clay particles are laminars, strongly bonded within themselves but only weakly bonded between one another. The laminar form of the particles tends to promote slip which is further assisted by the presence of water. Typically, clay contains water in two distinct ways – as water intimately linked to the clay particles by adsorption, and as free water. This is illustrated in Fig. 2.2. As the amount of water present in the system increases, so the plasticity – readiness to deform – of the clay increases; wet clay is very plastic and is often unsuitable for use in or beneath the pavement.

2.3 Tests for soils

In order that the response of a soil to changed circumstances can be predicted, the engineer often requires physical tests to be carried out. In the context of highway engineering the tests most generally used are those for particle size distribution as previously discussed, moisture content, compaction characterististics, California bearing ratio and liquid and plastic limits. Full details of these are given in BS 1377.

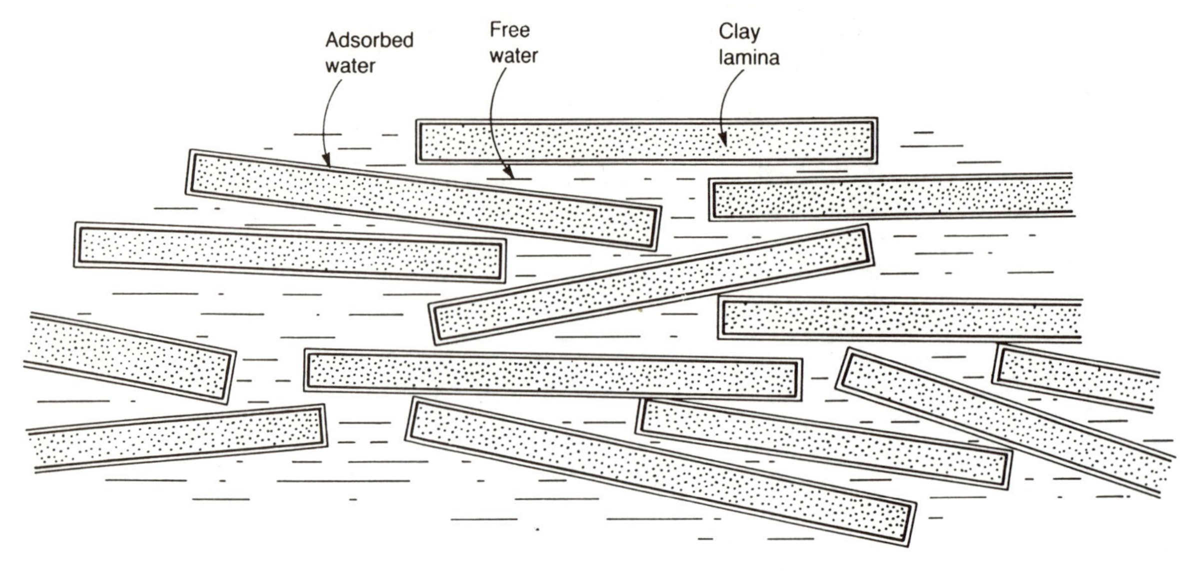

Figure 2.2 Clay/water system

2.3.1 Moisture content

The moisture content of a soil has an effect on its behaviour in that very dry cohesive soils are difficult to compact, very wet granular or cohesive soils lack stability and are difficult to compact, and the moisture content of a soil can influence chemical reactions taking place within it. Moisture content is defined as the ratio of the mass of water present in a body of soil to the mass of the dry soil particles. It is often expressed as a percentage and is measured by weighing a sample of the soil, drying it – usually in an oven at a temperature of 105 °C, and weighing it again. The mass of water is determined by subtraction.

2.3.2 Compaction

A soil's compaction characteristics are of interest when any area of fill is to be placed, including work either above or below the formation. The problem in the field is to ensure that in areas of fill the soil is sufficiently compacted to make it likely that only an insignificant amount of settlement will take place after construction. This propensity to settle is related to the density of the compacted soil. Density is in turn related to the nature of the soil, its moisture content and the effort expended in compaction. The moisture content and compactive effort on site must be right.

A standard test is available which enables the response of a soil to variations in moisture content and compactive effort to be studied. This test is fully described in BS 1377. The test equipment consists of a standard cylindrical mould into which the soil to be tested is introduced in layers, and a falling weight hammer which is used to impact a standard compactive effort to each layer of the soil. Moisture content is varied from sample to sample and measured, and the weight and hence the density of the known volume of soil

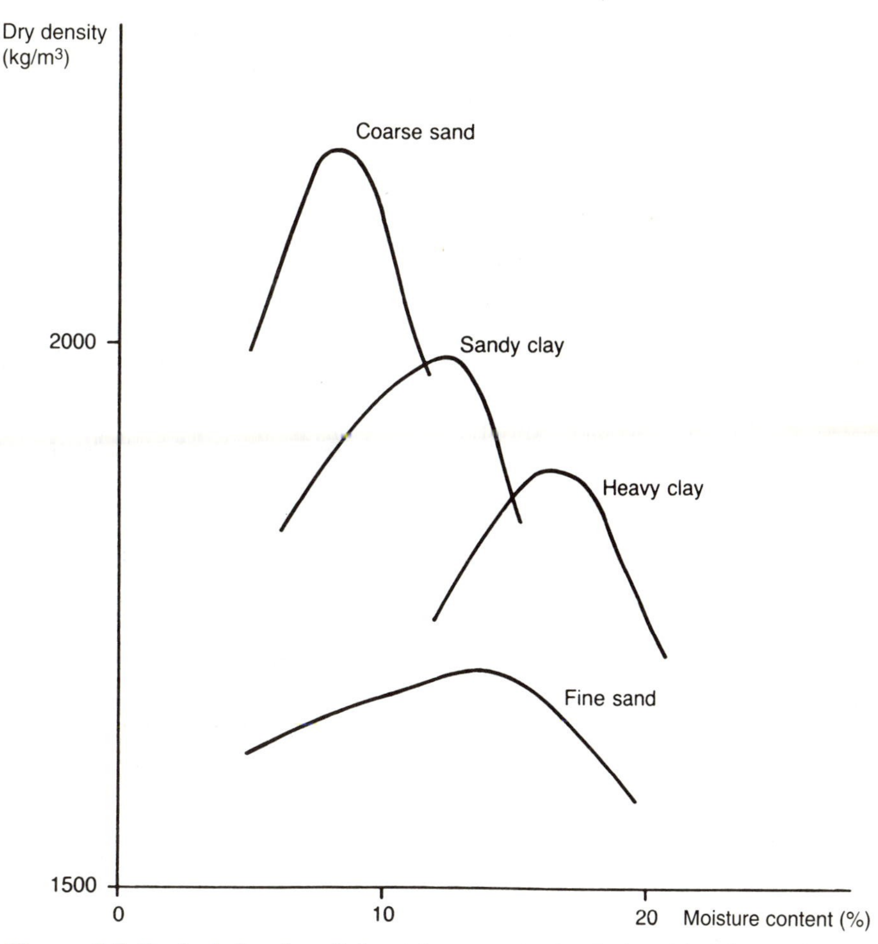

Figure 2.3 Typical dry density/moisture content relationships for various soil types

in the standard mould is recorded for each sample. It is thus possible to find how the dry density (the weight of soil solids in unit volume) of the soil varies with its moisture content, given a constant compactive effort. Typical results from such tests are shown in Fig. 2.3.

At low moisture contents the soil is stiff, unworkable and difficult to compress but as the moisture content is increased so the water acts as a lubricant between the soil particles, easing relative movement, and the soil becomes easier to compress. As further water is added to the mix the voids between the soil particles become almost completely filled with water whose effect is to keep the soil particles apart – and so the density falls. Increases in moisture content increase the proportion of voids present in the soil and further reduce its density.

The moisture content at which the maximum density is obtained is reported as the optimum moisture content. It is the aim of the construction process to control the moisture content of the soil to within a few percentage points of this optimum value and to achieve a density very closely approaching the maximum.

Note that different soils have different optimum moisture contents. Typical results are, for sand/gravel mixtures, 5–7 per cent; for sands, 8–10 per cent; and for silts and clays, their plastic limit.

In order that soil compaction may be adequately monitored on site it is necessary to measure the density of soil after compaction. This may be done in either of two general ways – by the removal of a small sample of soil and the calculation, measurement or deduction of the volume and mass of the sample (standard tests are available, for example in BS 1377); or by the use of a nuclear gauge.

A great disadvantage of density measurement by the extraction of a sample and measuring its mass and *in situ* volume is the time taken for each determination. The nuclear gauge offers considerable savings in this respect.

The essential features of a nuclear gauge are a radioactive source, means of detecting and measuring radiation which has passed from the source through the sample, and suitable protection for the operator. Two modes of operation are available and illustrated in Fig. 2.4; direct transmission, in which the source is inserted into a preformed hole in the sample material, the radiation passing through the material to the detector; or the backscatter method in which both the sample and the detector are placed at the surface, some radiation being reflected back by the sample to the detector. In either case, gamma rays from the source pass through the material and the amount of radiation reaching the detector is inversely related to the density of the sample. Nuclear gauges are available which provide direct readouts in units of material density. It is also possible to use some nuclear gauges for assessment of moisture content by the backscatter method.

All possible care should be taken in the handling of radioactive materials.

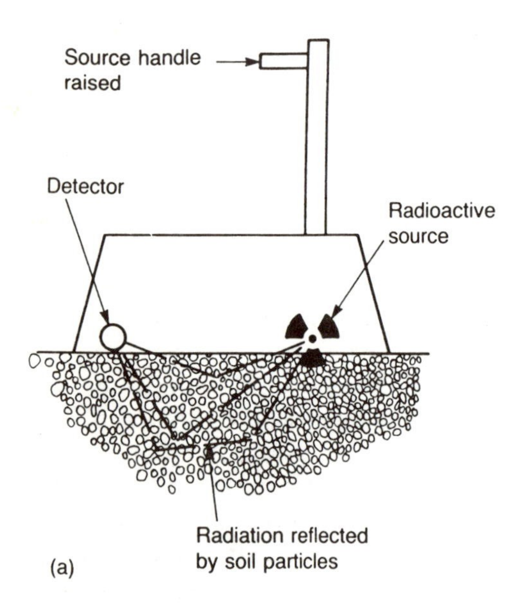

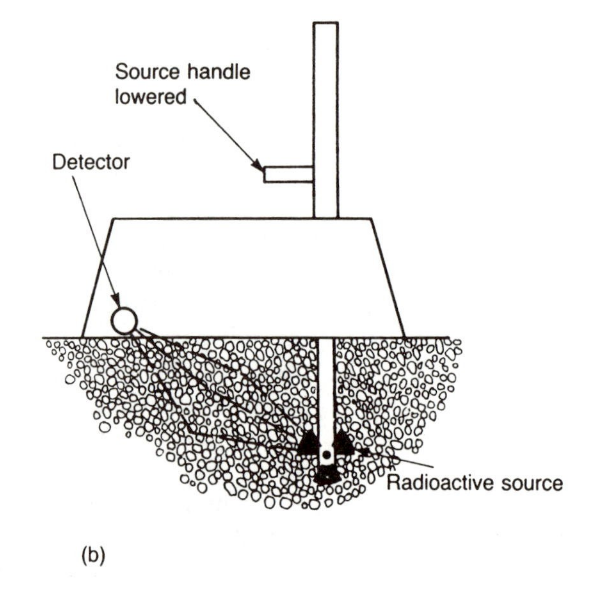

Figure 2.4 The nuclear gauge

2.3.3 Moisture condition test

The moisture condition test was devised[2] to enable the suitability of soils for use as engineering materials to be rapidly and reliably assessed in the laboratory or on site. It has the advantage of providing results quickly and is therefore suited to quality control in processes which are expected to modify the condition of a soil – such as lime stabilisation.

The test consists of applying standard blows of a falling weight to a sample of soil and monitoring the compactive effect which this has. The degree of compaction of the soil sample within the standard 100 mm diameter mould is measured by means of a vernier attached to the falling weight, read when the weight is resting of the top of the sample. Penetration is recorded together with the corresponding cumulative number of blows. The penetration caused by each blow progressively reduces as the test proceeds, owing to the increasing compaction of the sample. Penetration at any stage is compared with that achieved by four times as many blows, and this 'change in penetration' is plotted against the smaller number of blows as in Fig. 2.5. The moisture condition value (MCV) is defined as 10 times the common logarithm of the number of blows corresponding to a change in penetration of 5 mm.

Typical specification requirements are that soils for earthworks should have an MCV of not greater than 8, and that soils for lime stabilisation should have an MCV of 12.

2.3.4 California bearing ratio

When in service beneath a pavement, a soil is required to resist deformation which locally applied forces tend to cause. For example, a wheel load acting on top of a thin layer of unbound granular material overlying a stiff soil may cause a small displacement in that soil; the same wheel load applied to a similar layer of granular material which overlies a soft soil will tend to cause a much greater displacement in the soil. In the design of the lower layers of the pavement a knowledge of the likely response of the subgrade to such loading is valuable.

The California bearing ratio (CBR) test represents an attempt to quantify the behaviour of a soil in such circumstances. The test was developed in California in the 1930s and now forms the basis of the most widespread empirical pavement design method.

The test is an arbitrary one which does not attempt to measure directly any of the 'fundamental' properties of the soil sample. In essence it consists of driving a standard cylindrical plunger into the soil sample at a standard rate of penetration and measuring the resistance to penetration offered by the soil. This resistance is then compared with certain standard results, the ratio of the result for the soil to the standard result being reported as the CBR. The essential features of the test equipment are shown diagrammatically in Fig. 2.6.

In situ CBR tests are sometimes used instead of laboratory tests. Here the plunger load is usually applied from a lorry or other large vehicle, and the reference against which penetration is measured is achieved by a part of the apparatus bearing on the untested soil nearby. The surface of the soil to be

Soil: Heavy clay   Moisture content: 26.3 per cent

Number of blows of rammer 'n'	Penetration of rammer into mould (mm)	Change in penetration between 'n' and '4n' blows of rammer (mm)
1	41	33.5
2	57.5	33
3	67	33.5
4	74.5	26.5
6	84	17
8	90.5	10.5
12	100.5	0.5
16	101	
24	101	
32	101	
48		

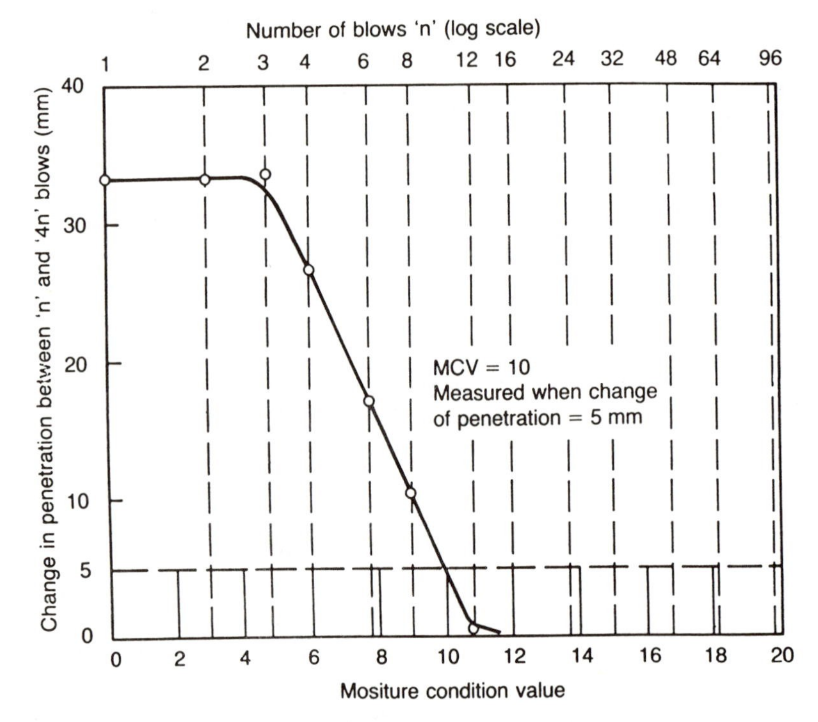

Figure 2.5 The determination of the moisture condition value of a sample of heavy clay

tested should be trimmed to accept the plunger uniformly across its full face, and also to accept any surcharge annular weights if these are used.

There are two main variables which can affect the test results, and the properties of the soil which the test quantifies. The moisture content of the soil can be very significant, particularly (but not exclusively) in the case of cohesive soils. It is for this reason that care must be taken in the control of moisture content during laboratory testing, and more particularly to ensure

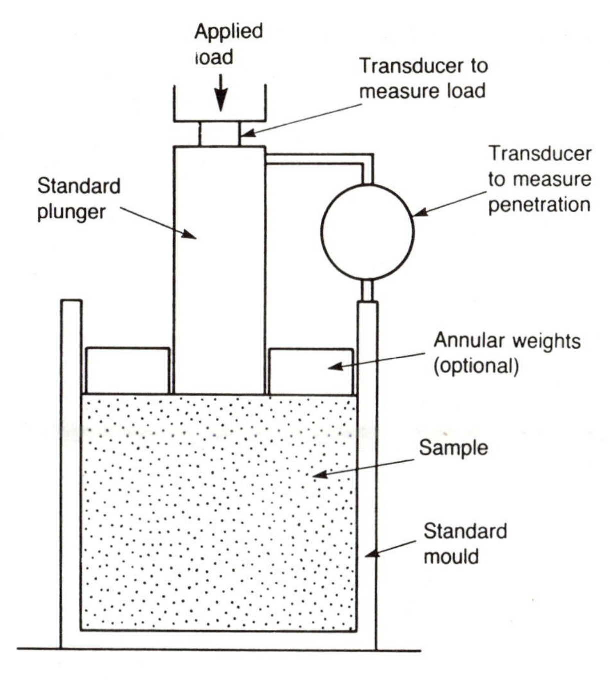

Figure 2.6 The California bearing ratio test

that the test conditions adequately represent those which will be experienced in the field.

The ability of a soil to resist deformation is also influenced by any surcharge acting on the soil. Provision may be made in the test for this phenomenon by applying annular weights around the plunger to impose a local surcharge on the soil, the intention being that this surcharge should equal that which will be imposed by the weight of the pavement which is expected to overlie the soil in service. As the surcharge increases, so will the effective CBR.

It is of course essential that any test used as the basis for a design method should be capable of giving reproducible results – that is, repeated tests by different individuals in different locations should produce the same result when carried out on the same soil. Unfortunately this is not the case for CBR tests on wet cohesive soils such as occur widely throughout the British Isles and elsewhere. For such soils whose CBR typically lies below about 5 per cent, test procedures cannot be sufficiently standardised to yield reproducible results, and for this reason the practice has arisen in such cases of estimating CBR values from other tests. These tests are those which determine further arbitrary properties of the soil, the liquid limit and the plastic limit.

2.3.5 Liquid and plastic limits

These properties of cohesive soils are used to estimate CBR values for use in the design of pavements.

The liquid limit of a soil is defined as the moisture content at which a soil passes from the plastic to the liquid state as defined by the liquid limit test.

Liquid limit may be determined in various standard ways; these are described in BS 1377 and elsewhere. The most convenient and frequently used test is that using the cone penetrometer. The tests consists of taking a sample of the soil passing a 425 μm sieve, mixing this thoroughly with water and placing it in a standard metal cup. A needle of standard shape and weight is then applied to the surface of the sample and is allowed to bear on it for five seconds. The penetration into the sample is recorded to the nearest 0.1 mm. The moisture content of the soil sample is determined. This process is repeated several times with different moisture contents and a graph of penetration versus moisture content is drawn. The moisture content corresponding to a cone penetration of 20 mm is obtained from this graph and reported, as a percentage to the nearest whole number, as the liquid limit of the soil obtained by the cone penetrometer method. This method is preferred as it yields the most reproducible results.

Plastic limit is also a property of cohesive soils and is defined as the moisture content at which a soil becomes too dry to be in a plastic condition as defined by the plastic limit test. The standard test consists of taking a 15 gram sample of the soil and mixing it with distilled water until it becomes plastic enough to be rolled into a ball. The material is then repeatedly rolled by hand into a thread of 3 mm diameter, the moisture content gradually being reduced by evaporation in the process. Eventually the soil becomes so dry that it crumbles when rolled into a 3 mm thread; this moisture content is reported as the plastic limit.

Thus both liquid and plastic limits are moisture contents, expressed as percentages by mass. Liquid limit of course always exceeds the magnitude of the plastic limit, and the difference between the two indicates the way in which the strength properties of the soil vary with changes in the moisture content. This numerical difference is known as the plasticity index (*PI*) thus:

$$\text{Plasticity Index} = \text{Liquid Limit} - \text{Plastic Limit}$$

and has been found to bear an empirical relationship to the CBR of soft cohesive soils. Since the liquid and plastic limits tests have been found to give superior repeatability to the CBR test, it is these former which are used in the pavement design process. We have seen that CBR depends at least in part on the service conditions – moisture content and surcharge – and Table 2.2 represents the best available estimate of CBR from a knowledge of plasticity index and of service conditions. Note that in this table:

- A high water table is 300 mm below the formation;
- A low water table is 1000 mm or more below the formation;
- A thin pavement is 300 mm thick;
- A thick pavement is 1200 mm deep including capping;
- Poor conditions are when the lowest layer of the pavement is laid on weak soil in heavy rain
- Average conditions are when the formation is protected promptly during variable weather;
- Good conditions are when the soil is drier than its likely service condition throughout construction.

Table 2.2 Typical CBR values for various soil types and conditions[3]

Soil type	PI	*High water table*						*Low water table*					
		Poor		*Average*		*Good*		*Poor*		*Average*		*Good*	
		A	*B*	*A*	*B*	*A*	*B*	*A*	*B*	*A*	*B*	*A*	*B*
Heavy clay	70	1.5	2	2	2	2	2	1.5	2	2	2	2	2.5
	60	1.5	2	2	2	2	2.5	1.5	2	2	2	2	2.5
	50	1.5	2	2	2.5	2	2.5	2	2	2	2.5	2	2.5
	40	2	2.5	2.5	3	2.5	3	2.5	2.5	3	3	3	3.5
Silty clay	30	2.5	3.5	3	4	3.5	5	3	3.5	4	4	4	6
Sandy clay	20	2.5	4	4	5	4.5	7	3	4	5	6	6	8
	10	1.5	3.5	3	6	3.5	7	2.5	4	4.5	7	6	>8
Silt		1	1	1	1	2	2	1	1	2	2	2	2
Sand													
poorly graded								20					
well graded								40					
sandy gravel								60					

Notes:
PI = plasticity index
Poor, Average, Good refer to construction conditions
A = thin pavement construction
B = thick pavement construction

2.4 Disadvantageous soils

Naturally occurring soils can have a wide range of chemical and physical properties, the great majority of which are of little corcern to the highway engineer. However, attention must be paid to two properties of the soil which can seriously affect the stability or durability of the pavement. These are the probable response of the soil to frost, and the presence in the soil of chemicals which will degrade stabilising materials within the pavement.

2.4.1 Frost susceptibility

The reader may be familiar with the notion of damage to surfaces exposed to frost; small fissures are penetrated by water which then freezes and expands, causing enlargement of the fissures. The cycle is repeated, leading to widespread disruption of the surface. This type of damage can be avoided in a pavement by keeping the surface free of fissures.

A more serious and deep-seated cause of failure can arise when the subgrade below the pavement freezes. In certain types of soil ice lenses form, accompanied by the upward capillary movement of moisture which also freezes, leading to the formation of ice lenses up to 25 mm thick. These lenses cause vertical movement at the surface roughly equal to their own thickness, and upon thawing greatly increase the moisture content of the soil in which they are formed – further tending to induce failure of the pavement. Soils which exhibit this tendency are said to be frost susceptible and include:

- Cohesive soils with a plasticity index less than 15 per cent (well drained soils) or 20 per cent (poorly drained soils – within 600 mm of the water table);

- Limestone gravels with aggregate saturation moisture content greater than 2 per cent;
- Crushed chalk;
- Limestones – hard limestones with more than 2 per cent aggregate saturation moisture content;
  – Oolitic and Magnesian limestones with more than 3 per cent aggregate saturation moisture content;
- Burnt colliery shale;
- Pulverised fuel ash with more than 40 per cent passing a 200 μm sieve.

The designer should therefore avoid placing any of these soils in such a position as will allow freezing to take place. In the UK climatic conditions are such that if no frost susceptible material is placed within 450 mm of the surface of the pavement there will be no practical risk of the material freezing and heaving. In areas where frost susceptible soils will lie immediately below the formation this minimum pavement thickness should be provided in non-susceptible materials irrespective of other considerations. Elsewhere in the world the likely intensity of frost may be expected to differ and the minimum thickness of non-frost susceptible material should be varied accordingly. There may also be benefits to be obtained from lowering the water table, thus reducing the ability of groundwater to move up into the frosty layers.

A standard test is available[4,5] enabling the frost susceptibility of a soil to be assessed. The essential feature of the test is that a sample of the soil is placed with its lower end in water at a standard temperature of 20 °C and is frozen at the top; the amount of heave is monitored during a standard time and the material is regarded as being frost susceptible if this heave exceeds 12.7 mm in these conditions.

2.4.2 Sulphates

It is sometimes the case that aggressive chemicals are present in small but significant quantities in soil or groundwater. Sulphates are found in some clays, in sea water and in some groundwater and can cause deterioration of concrete bound with ordinary cement. In cases where the sulphate content of the soil is found by chemical analysis to exceed 0.2 per cent or the acidity of the soil to be pH <6.5 the engineer should ensure appropriate precautions are taken, such as the use of a suitable sulphate resisting cement.[6]

2.5 Bitumens

Simple pavements of unbound graded stone were once commonplace – and still are in many parts of the world – but when subjected to more than the most nominal amounts of traffic they suffer from their lack of surface stability. This causes large amounts of dust, loss of material from the surface, rutting, and penetration of water into the pavement and the soil beneath. An early technique used to combat this lack of stability was the spraying of tar onto the surface to bind and stabilise the roadstone. From this simple beginning there have developed a wide range of applications of bituminous materials in pavement construction.

There are two very general types of bituminous binder. Tar is obtained from coal, usually as a by-product of the manufacture of coke or the production of coal gas. With the increased availability of natural gas the availability of tar has reduced.

Bitumen is obtained from crude oil. The reader may be familiar with the concept of the fractionation of oil – a process in which crude oil is heated and the various groups of hydrocarbons which form this mixture of materials are driven off. The most volatile (petrol) leave the system first and as the temperature rises so kerosene, gas oil and other heavier fractions leave the mixture until a residue of bitumen remains. This residue is a mixture of many different and complex materials, and its composition and properties will depend on its source and manufacture. Bitumen is used in a variety of ways to stabilise and strengthen roadstone, particularly in the highly stressed upper layers of the pavement.

Properties of bitumens which are of interest to the highway engineer are their ability to resist deformation, their response to changes in temperature, and their solubility in other hydrocarbons.

2.5.1 Resistance to deformation and response to temperature changes

When stress is applied to any solid body, deformation of that body will result. In the case of most solids the deformation will either be elastic (that is, reversible and proportional to the magnitude of the applied force) or permanent. When stress is applied to a fluid that fluid will flow to a greater or lesser extent, depending upon the applied stress and the fluid's viscosity (defined as the resistance of a fluid to flow; the unit of viscosity is the poise, defined as the resistance in dynes presented by the liquid to a surface of area 1 square centimetre moving at 1 centimetre per second at a distance of 1 centimetre from another fixed surface, all submerged in the fluid). The behaviour of a bitumen in response to applied stress is complex and depends not only on the size of the applied force but also upon the duration of its application and the temperature of the bitumen. It is therefore unreasonable to expect to be able to predict bitumen response to loading from a simple assumption of elastic behaviour.

This difficulty has long been recognised and was addressed by van der Poel[7] who proposed the term 'stiffness' to represent the ratio of stress to strain for bitumen at a particular combination of temperature and duration of loading. The way in which stiffness may vary with time and temperature for a typical bitumen is illustrated in Fig. 2.7. At low temperatures and when loads are applied briefly elastic behaviour may be expected; where the temperature rises or the load is applied for a long time the stiffness of the bitumen falls and deformation may be expected. This characteristic of bitumens is very important to the highway engineer and has been the subject of much research.

Two tests are generally used to assess the viscosity of a bitumen and thereby deduce its likely response to changes in temperature. These are the penetration test and the softening point test.

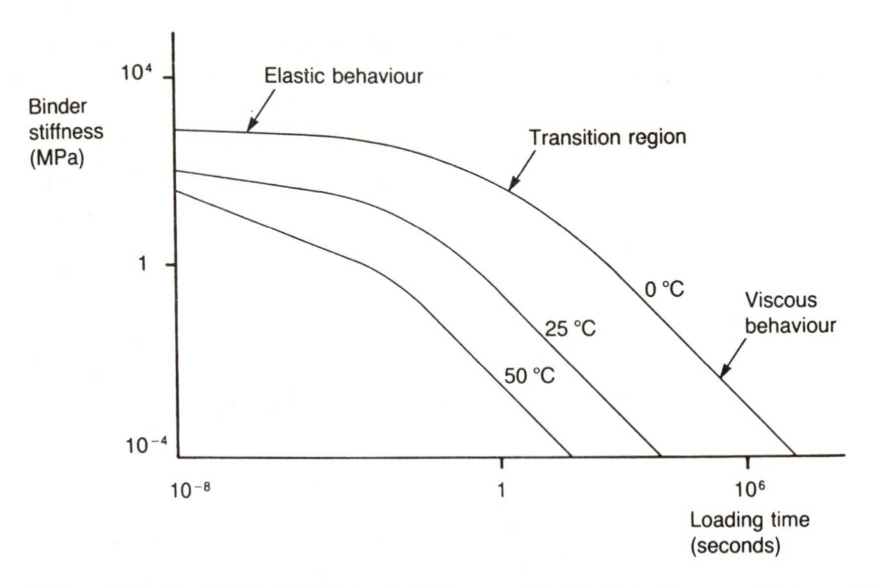

Figure 2.7 Variation of bitumen stiffness with loading time and temperature

The penetration test consists of applying a standard steel needle vertically to the top of a sample of the bitumen at a standard temperature of 25 °C, applying a standard load of 100 g and measuring how far the needle penetrates the bitumen during a five second period. The result is reported in units of 0.1 mm; thus a needle movement of 5 mm in the standard conditions would be reported as a penetration of 50; the bitumen might be referred to as a '50 pen binder'. Clearly the lower the penetration, the harder the bitumen tested, and vice versa. This test is often used as an indication of the nature of certain grades of bitumen.

The softening point test takes a sample of binder which has previously been cast inside a metal ring of 15 mm internal diameter and 6 mm height. This sample and ring are placed in a water bath initially at a temperature of 5 °C. The ring is supported in such a way that a clear space 25 mm high exists below the sample. A 10 mm steel ball is placed on the sample and the temperature of the water bath, and therefore of the sample, is raised at a uniform rate of 5 °C each minute. With increasing temperature the bitumen softens and sags under the weight of the ball. The temperature at which the sagging binder first reaches the bottom of the 25 mm gap below its initial position is reported as the softening point. This is also known as the ring and ball test and is illustrated in Fig. 2.8.

Both of these tests give repeatable results if carefully carried out but both are somewhat arbitrary in respect of the properties of the material which they measure and in this reflect their pragmatic origins.

The test results enable the material supplier and the engineer to predict the temperatures necessary to obtain sufficient fluidity in the mixed material for it to be used effectively. There are very different requirements of the viscosity of a binder which include a relative low viscosity (up to about 2 poise) required at the time of mixing a bituminous roadstone, to ensure that the binder will

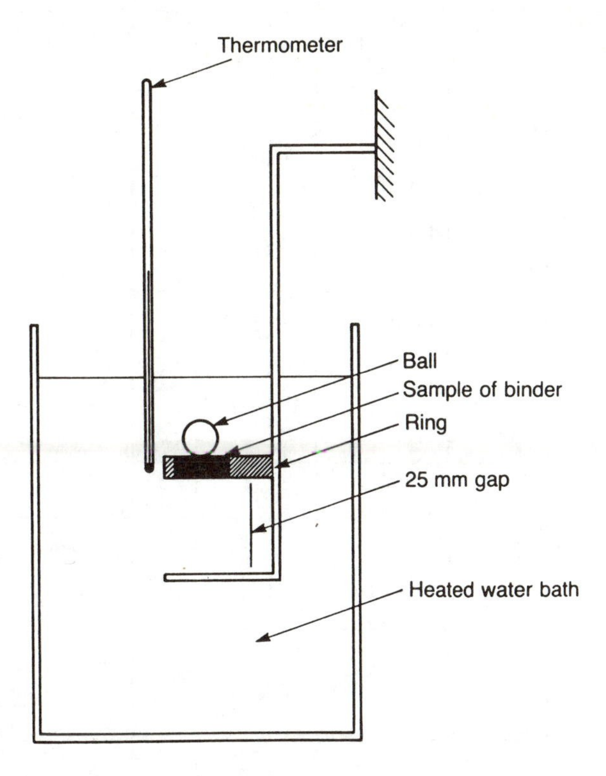

Figure 2.8 The softening point test

properly flow over and coat the surface of the aggregate; an increased viscosity of up to 20 poise at the time of laying, in order that the mixed material may be sufficiently fluid and yet not segregate; and when the material is compacted a viscosity of not more than 300 poise is appropriate so that a normal compactive effort will achieve the desired result of a dense, relatively void-free finished layer.

Figure 2.9[8] represents a method of estimating the temperature susceptibility of a bitumen from results of penetration and softening point tests, and also enables appropriate minimum temperatures for mixing, laying and rolling the material to be deduced. The chart may be used by:

- Idenfitying a point on the 25 °C ordinate which corresponds to the penetration of the bitumen;
- Identifying a point on the dotted line (SP) which corresponds to the softening point of the bitumen;
- Constructing a straight line through these two points and projecting it as necessary;
- Reading off the appropriate temperatures.

The gradient of the constructed line is indicative of the bitumen's sensitivity to changes in temperature; a steeply sloping line will correspond to a material whose stiffness is very temperature dependent while a nearly horizontal line

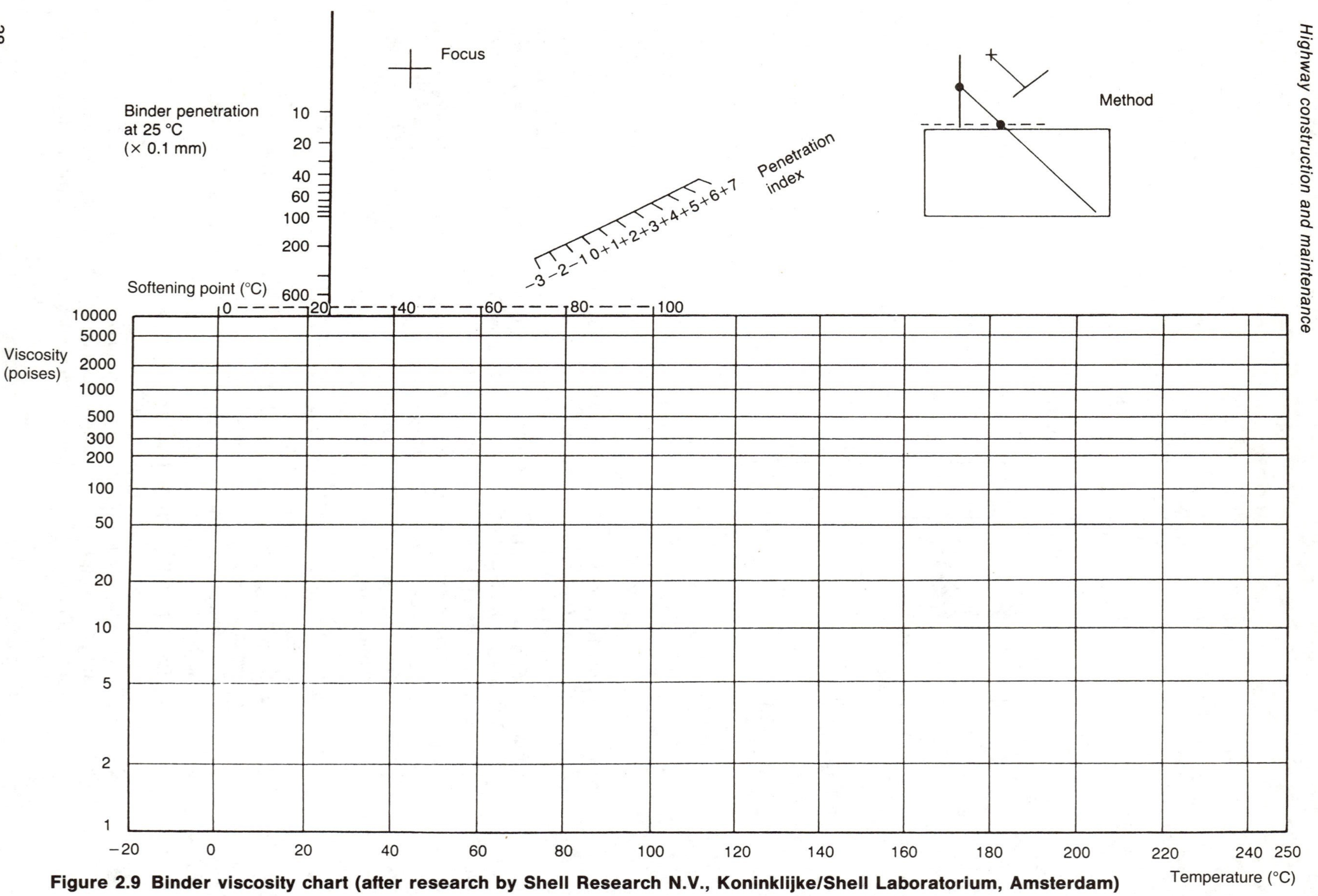

Figure 2.9 Binder viscosity chart (after research by Shell Research N.V., Koninklijke/Shell Laboratorium, Amsterdam)

on the diagram would indicate a bitumen whose stiffness is almost independent of temperature. The term 'penetration index' (PI) has been coined to provide a measure of this temperature susceptibility. PI may be obtained from Fig. 2.9 by drawing a line parallel to that previously constructed through the focus to the PI scale and reading off the appropriate value therefrom; or from the equation[9]

$$\mathrm{PI} = \frac{1951 - 500 \log P - 20SP}{50 \log P - SP - 120.1}$$

in which P is the penetration and SP is the softening point.

Penetration index is of value in the analytical design of bituminous pavements.

These rheological properties of bitumens are modified by heating and laying the material. Generally one may expect a hardening to take place as a result; if the initial penetration is represented by P_i and the penetration of material from the same source recovered from a pavement by P_r then the following approximate relationship exists:

$$P_r = 0.65P_i$$

It will be readily appreciated that binders from different sources will not necessarily have the same viscosity characteristics. It is important that the binder in a bituminous pavement, particularly one of rolled asphalt in which the binder makes the most significant contribution to the strength of the pavement, should retain certain properties in terms of its stiffness throughout the temperature range likely to be experienced in service. The binder should remain slightly plastic; too hard a binder may result in cracking of the layer, or possibly in the loss of chippings from the surface, while too soft a binder will of course lead to deformation. Unfortunately bitumens obtained from the distillation of crude oil (sometimes known as residual bitumens, RBs) do not exhibit the wide plasticity range that one might seek for the most heavily trafficked pavements; their *PI* is typically rather less than zero.

The situation may be improved somewhat by blending residual bitumen with a more stable material. A 1:1 mixture with the naturally occurring Trinidad Lake Asphalt will effect some improvement, particularly in reducing the flow of the material as determined in the Marshall test, but the most successful example of the blending of binders is that of the polymer modified bitumens which, as the name suggests, consist of a blend of suitable synthetic polymers, Trinidad Lake Asphalt and RB. These modified binders are commonly available in two grades, one for stone contents of less than 50 per cent and one for more. Typical properties of various binders are illustrated in Table 2.3.

Yet another additive which has been successfully included in bitumen is epoxy resin. Such resins are generally used by mixing two fluid components which react chemically with one another to form a durable solid material. The rate at which the reaction proceeds depends on the ambient temperature and, where the epoxy is extended by the addition of a third material, the concentration of the two epoxy components. The presence of an epoxy resin in a binder imparts to the mixture high stability, stiffness and fatigue strength together with the ability to withstand the solvent action of kerosene and

Table 2.3 Typical bituminous binder properties

Binder	Penetration at 25 °C	Softening point	Penetration index
50 pen residual bitumen	50	52	−0.6
35 pen residual bitumen	35	57	−0.2
Pitch-bitumen	46	53	−0.5
Polymer modified type 1	47	80	+4.0
Polymer modified type 2	90	75	+5.5
Lake Asphalt	2	95	+0.5

petroleum. Asphalt incorporating this binder has been marketed for use in heavy duty surfacings such as at container terminals and airfield aprons but the high cost has often proved to be a deterrent to its use – particularly since the use of block paving for such applications became widespread.

2.5.2 The need to modify binder properties

In some cases it is quite convenient to accept the requirements of blended bitumens for high handling temperatures; in some cases it is not and we therefore seek ways of modifying the stiffness of bitumens during the construction stage without greatly affecting their properties in the long term. There are two general ways of achieving this.

2.5.2.1 Cutback bitumens

Bitumens obtained from the fractional distillation of crude oil may be blended with some of the more volatile fractions to produce a solution which will have a viscosity much lower than that of the pen grade bitumen and will therefore be fluid at much lower temperatures; however upon exposure to the air the volatile fraction will gradually evaporate to leave the bitumen behind. This process is known as curing and the speed of curing will of course depend on the nature of the solvent to which the bitumen is added. Bitumens which have been modified in this way are known as cutbacks, and are typically available in three general grades: slow-, medium- and fast-curing. Slow-curing cutbacks consist of a mixture of bitumen and oils of low volatility and therefore only gradually make a contribution to the strength of a coated roadstone; their commonest uses are in stabilising the surfaces of otherwise unbound roads, particularly to reduce dust; and in fine cold asphalt. Fast-curing cutbacks are produced by dissolving bitumen in petroleum or similar materials and suffer from the disadvantage that they are dangerously inflammable; they are as a result little used. The commonest group of cutbacks contain the medium-curing binders, in which the bitumen is dissolved in kerosene.

Cutback bitumens are classified in terms of their viscosity and of the penetration of the non-volatile residue. The viscosity of a cutback is measured with the standard tar viscometer. The test consists of measuring the time taken for a standard volume of binder to flow through a standard orifice in the bottom of a standard container, all at a temperature of 40 °C. The time taken, in seconds, is reported as the viscosity (STV). Three grades of cutback are commonly available in the UK, having viscosities of 50, 100 and 200 seconds STV respectively (depending on the proportion of volatile fractions present), and all having residue penetrations of between 100 and 350. Cutback bitumens of this type are used in open textured coated macadams and for surface dressing.

2.5.2.2 Bituminous emulsions

An alternative method of making bitumens easier to handle is by the formation of emulsions, in which minute particles of bitumen are suspended in water. The particles are coated with a film of material which bears an ionic charge, and since all bitumen particles in an emulsion are coated with the same material bearing the same ionic charge the globules are mutually repulsive and therefore do not merge when they come into contact. The ionic charge may be positive, as in cationic emulsions, or more commonly negative, as in anionic emulsions.

Bitumen emulsions are classified in terms of their stability or rate of break, and the proportion of the emulsion which is bitumen.

Rate of break depends on the composition of the emulsion, the rate at which the water evaporates which in turn depends on the ambient conditions and the method of application, the porosity of the surface to which the emulsion is applied, mechanical disturbance of the emulsion by rolling and the action of traffic, and the chemical and physical influence of the surface to which the emulsion is applied. Cationic emulsions tend to break more rapidly than anionic emulsions. Bituminous emulsions are classified from 1 to 4 in terms of their stability, class 1 being the most stable.

Emulsions are readily available in the UK in a range of binder contents from 40 to 70 per cent. They may of course readily be diluted in water. In the UK the standard nomenclature of bitumen emulsions indicates the polarity of the ionic charge, the stability of the emulsion, and in many cases its binder content. Thus A2–50 represents an anionic emulsion of class 2 stability, containing 50 per cent bitumen; K1–70 a more stable cationic emulsion with 70 per cent bitumen content.

2.5.2.3 Other additives

The characteristics of bitumens may be improved by means of various other techniques.

Pitch, when added to bitumen, has the effect of increasing the rate at which the binder oxidises on exposure to the atmosphere. One of the effects of oxidation is that the binder loses flexibility and therefore becomes harder and more susceptible to the abrasive effects of traffic. An oxidised binder is thus likely to wear away in preference to particles of coarse aggregate set in the surface, allowing the chippings to protrude from the surface and improve the skidding resistance. Pitch–bitumen mixtures should contain of the order of ten per cent pitch by mass. Note that it is possible to modify the penetration of a binder at the refinery by blowing air through it, thus encouraging oxidation.

Rubber may be added to bitumen in various ways – as liquid latex, for example, or in powder form. The effect of adding about four per cent of natural rubber by weight of binder is, unsurprisingly, to improve the elastic properties of the binder. This is advantageous where a bituminous surfacing may be otherwise expected to crack – as for example when laid over an old jointed concrete pavement, where reflection cracking may be expected over the movement joints. Tests[10] have suggested that rubberised asphalt is effective in these circumstances.

Sulphur may be used as a partial substitute for bitumen in some asphalts, and experimental work has been carried out in this field in the US. The economic benefits of using sulphur extended asphalt (SEA), whose binder consists of such a mixture, depend on the relative costs of the two materials. In the UK there is currently little cost benefit to be obtained from the use of sulphur, and SEA is therefore not used.

Sulphur provides the benefit of a strong and durable material. However its use should be approached with caution since at temperatures above 140 °C sulphur will take up water to yield the unpleasant gases sulphur dioxide and hydrogen sulphide and the consequent environmental drawbacks during the laying process militate against the use of sulphur in urban areas.

Foamed bitumen is formed when about 2 per cent of water is added to bitumen at mixing temperatures (of the order of 170 °C). The water rapidly boils and forms a foam in the molten binder. The properties of the foam are such that a volumetric increase of between 8 and 15 times ideally takes place – depending on the volume of water added – and the foam has a half life – time taken for its volume to reduce by one half – of about forty seconds. It has also been found that foamed bitumens offer a very effective and efficient way of coating aggregate, and that foamed asphalt has the advantage over pen grade bitumens that it can be used at a lower mixing temperature and over bituminous emulsions in that long curing times are not needed. Particular advantages have been found in the mixing of foamed asphalt with cold aggregates such as in the recycling of road materials (see Chapter 9) or in soil stabilisation.

2.6 Cement and lime

These materials are similar in appearance and in some functions but to the highway engineer they are significantly different.

2.6.1 Cement

The main property of cement is that it is capable of acting as a binder, forming a rigid matrix in which particles of aggregate may be set.

Cement is produced from calciferous material, such as chalk or limestone, and materials such as clay which are rich in silica and alumina. These are reduced to powder form by milling, mixed in the appropriate proportions and heated to such a temperature that the material sinters, forming a coarse clinker which when cooled and ground is the basic ingredient of ordinary Portland cement (OPC). This material is a mixture of calcium aluminate and calcium silicate, both of which react with water to form the familiar stone-like mass.

It was the resemblance of this hydrated material to Portland stone, a high-quality limestone, which gave the material its generic name. Since the cement reacts with water, it is important to ensure that sufficient water is present at the required time to enable the reaction to continue to completion, and that this water remains present while it is needed. If the mix is allowed to dry out while curing is taking place, weakness will result. If on the other hand too much water is allowed into the mix then the strength of the hydrated cement mass will again be impaired because the excess water will not be used up in the reaction and will remain to form voids in the hardened material, thus reducing its strength. The relationship between the water/cement ratio and the strength of the resulting hardened cement mass is illustrated in Fig. 2.10.

In order that the water initially in the mix may continue to be available it is often important to ensure that the water does not leave the system by, for example, evaporation; the highway engineer deals with large thin slabs in exposed locations and is often very concerned with this point. It is also of vital importance that the material does not freeze while setting since not only does this deny the availability of free water but also the volume changes involved can disrupt the material and cause irreparable loss of strength. Although heat is generated in the reactions which take place in curing, this is not sufficiently intense to be of much concern to the highway engineer because of the large surface of the slabs.

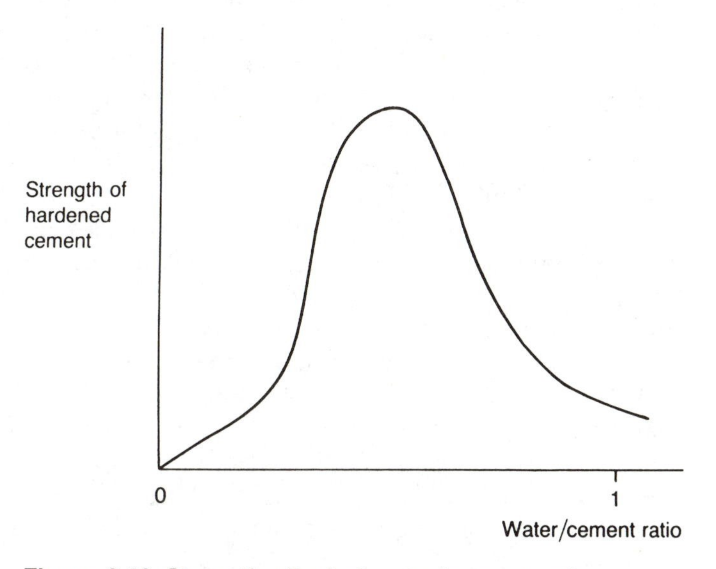

Figure 2.10 Strength effect of water/cement ratio

A characteristic of cement which is of great concern in highway engineering is the shrinkage which occurs in the hardening material, during and after curing. As curing proceeds, so some water is adsorbed within the cured material, causing the cement particles to swell. If the set cement is placed in an environment in which water tends to leave the system – such as relatively dry air – then some of the adsorbed water will be lost, the cement matrix will lose volume and the cementitious mass will shrink. Shrinkage is measured in units of strain; for mixtures of cement and water only – a cement grout – shrinkage of up to 4000 microstrain (4000×10^{-6} metres per metre) have been observed.[11]

Admixtures are available to modify the properties of the cement or of the concrete of which it forms a part. Many admixtures are available, to serve a wide range of purposes. Those which are most likely to be of interest to the highway engineer include air-entraining, colouring, and cement replacement.

Air-entraining agents consist of organic materials such as resins, fats and oils. They have the effect of producing in the wet cementitious mix a finely divided foam of very small, stable bubbles. These bubbles, when dispersed throughout the mix, have a lubricating effect, allowing the mix to flow more easily – a property which is often wanted when the concrete is being placed. Furthermore, it has been found that air-entraining concrete has a greater resistance to frost damage than does untreated concrete and is therefore better suited to form the exposed surface of a pavement. On the other hand the inclusion of entrained air will of course increase the proportion of voids in the concrete, and reduce its strength. This may to some extent be offset since the air, by increasing the fluidity ('workability') of the mix, may allow a reduced water content – water often being added to improve workability – which will in itself tend to reduce the proportion of voids in the material. Air-entraining agents are often used in the upper layers of concrete pavements. Sometimes the natural colour of the concrete may not be acceptable. Additives are available which will colour the mix to one of many subdued colours. The additives are widely used on concrete block paving (Chapter 6) where the concrete is mixed under carefully controlled uniform conditions. It is possible to colour concrete cast *in situ* by the addition of suitable colouring agents but this is very unusual for highway projects, where the applications for coloured concrete are not so large as to make worthwhile the use of sufficiently rigid quality control to ensure uniformity of colour.

An increasing awareness is developing of the harmful effects of de-icing salts on steel reinforcement in concrete, particularly in highway structures, and in reinforced concrete carriageways. One way of countering this is to replace some of the cement in the mix with some other suitable material so that the ability of the salt to penetrate the concrete is reduced, thus increasing the time taken before corrosion of the reinforcement becomes significant. Suitable materials are pulverised fuel ash (pfa) or ground granulated blastfurnace slag (ggbfs). Normal replacement levels are of the order of 25–30 per cent of the cement normally required in a mix, which should increase the protection afforded to the reinforcement by a factor of between four and six.[12]

2.6.2 Lime

Like cement, lime can act as a binder, forming a hydrated matrix in which aggregate may be set, and it was used in this way by the ancients for centuries before the invention of cement. Lime is however a poor substitute for cement and is therefore not used as a binding and adhering agent in modern civil engineering. The major use of lime in the highway is as a stabiliser of clays; its use is discussed in Chapter 3.

Lime is available in two chemical forms. Calcium oxide is produced commercially by burning limestone in kilns. It reacts with water, the violence of the reaction giving it the name of quicklime ('quick' having its archaic meaning of 'living') and being accompanied by the evolution of much heat. This reaction leads to the formation of slaked ('satisfied') or hydrated lime, a material of less aggressive characteristics but which is still of value in highway construction.

Quicklime is provided as a solid in lump (particle size generally exceeding 20 mm), granular (6–20 mm) or fine (less than 6 mm) form. Many users prefer the coarser sizes since there is less chance of the quicklime being blown by the wind onto unprepared people or property near the works. Quicklime is a material which should be handled with care because of the violence of its reaction with water. It should therefore be kept dry before use, stored away from inflammable materials and handled carefully; operatives should wear protective clothing, goggles and dust masks. Despite these disadvantages, quicklime is often used for clay stabilisation for the very reason of its great affinity for water. Slaked or hydrated lime is generally delivered to site in powder form but may be applied either dry or as a slurry, mixed with two party by weight of water.

2.6.2.1 The stabilising action of lime on clays

Natural clay consists of laminar particles which slide freely on one another and whose ability to slide is aided by the lubricating action of free water present in the system and of water molecules which become attached to the surface of the clay particles. This is shown in Fig. 2.2. Remove the water, and the plasticity of the clay is enormously reduced. Lowering the water table can have the effect of reducing the amount of free water, but cannot remove the water which is chemically linked to the clay; for the best results, a more positive approach is needed.

If we mix quicklime into clay a remarkable change occurs. Firstly, the quickline reacts with the free water in an exothermic reaction to form slaked lime. The heat generated is sufficient to cause further water loss by evaporation. The slaked lime then reacts with the clay particles, drawing off the molecular water and causing the clay particles to flocculate. The effect of this initial reaction, taking only hours to come into effect, is to change the nature of the soil to a relatively coarse-grained, free draining medium of much lower plasticity. In the following days further reaction takes place, cementing the flocculated clay particles into larger granules. Because the cementitious effect is within the particles rather than between them, lime-stabilised clay can be reworked at any time without detrimental effect.

Typically, the quantity of quicklime required to achieve worthwhile results is of the order of 2–3 per cent by dry weight of soil. Increases in CBR from less than 5 per cent to more than 50 per cent in a few days are by no means uncommon.

2.7 Aggregates for bound materials

As before, bound materials here include both bituminous and cementitious materials used in the pavement. Aggregates for these may be natural gravels, crushed rock, or from artificial sources. Natural gravels and crushed rock have already been discussed in general terms.

Synthetic aggregates are generally obtained from the by-products of large-scale chemical processes, such as the burning of domestic refuse or the production of iron.

Blast-furnace slag is a by-product of the production of iron and is widely used in bituminous materials for highway use. It consists of an amalgam of limestone with coke ash and aluminous and silicaceous residues of iron ore. Air cooled slag takes on the general appearance of an igneous rock and because of its high initial lime content is ready for use only after stockpiling and weathering for a period of up to about a year.

2.7.1 Desirable properties

Strength, response to abrasion and colour can be of concern here.

The strength of an aggregate – its ability to support direct compressive or tensile forces – is of interest in a concrete mix, since the strength of the material as a whole can be affected, and in a bituminous mix, since a weak aggregate may be crushed during compaction. Stresses which arise in the pavement are not so high that quantitative tests are necessary, however, and it is generally enough to ensure that the aggregate is of a type known to be acceptable – for example, suitable types of rock include basalt, gabbro, granite, gritstone, hornfels, limestones, porphyry and quartzite. Flint is acceptable in the form of a natural gravel only, whereas the others may be crushed or natural gravel. Air cooled blast-furnace slag can also be suitable; BS 1047[13] gives guidance.

Colour is of course of interest only when it shows – that is, when the aggregate is exposed at the surface of the pavement. In the top layer of the pavement the changes in colour which are available are generally somewhat subtle and are thus unlikely to be of great interest in themselves as means of modifying the final appearance of the work but it should be noted that abrupt changes in subtly differing shades can mar an otherwise satisfactory pavement. As an extreme example, consider the effect of taking delivery of and using chippings supplied from two alternative sources, one yielding grey and one red; the resulting chequer board effect would be most unpleasing and would look even worse when reinstatments were made using yet another colour of chipping. Providing that extremes of colour are avoided, little difficulty arises in practice from this aspect of the work.

The aggregate characteristic which most often concerns the engineer is its ability to respond favourably to the abrasive effect of traffic. The requirement here is that chippings exposed at the surface should help to provide adequate skidding resistance.

In wet conditions vehicle wheels most often tend to lose their grip on the road surface, and so the designer seeks to promote high skidding resistance in the surface. A successful design will create the situation where the wheel bears on a slightly roughened surface, promoting the efficient transfer or stresses from wheel to road, and where the wheel/road interface is more or less dry, even in wet weather. Motor vehicle technology has made many contributions to means of resolving this problem, particularly in the field of tyre design. The highway engineer's contribution is to ensure that suitable aggregate is exposed throughout the surface in such a way as to be free-draining.

A well drained interface between wheel and road is best achieved by ensuring that chippings or other features are provided which project a sufficient distance above the general surface of the pavement to stand proud of the film of surface water that develops in times of heavy rain. This requirement demands a degree of roughness on a large scale, known as the macrotexture; high macrotexture provides drainage paths for the removal of water from the wheel/road interface and enables the wheel to effectively penetrate thin films of water. In practical terms, macrotexture is initially a function of the rate of spread of chippings on a bituminous surface, and their degree of embedment; in the long term these factors are still important but are joined by the ability of the aggregate to resist abrasion. This may be assessed by means of the aggregate abrasion value (AAV) test.

This test consists in essence of finding the percentage loss of mass of a sample of aggregate when subjected to standard conditions of wear for a standard period. Clearly, the lower the AAV, the less material will have been lost in the test and therefore the higher will be the expected resistance to abrasion. It is however desirable to provide for a very small degree of wear in surface chippings so that they are not left unsupported by the general surface of the road as this itself is very slowly worn away. The choice of a suitable AAV will of course depend on the amount of wear to which the road is expected to be subjected; the more traffic, the more wear and the lower the acceptable AAV. Typical requirements[14] are shown in Table 2.4.

Table 2.4 Recommended maximum aggregate abrasion values

Traffic in commercial vehicles per lane per day	*Under 250*	*Up to 1000*	*Up to 1750*	*Up to 2500*	*Up to 3250*	*Over 3250*
Maximum AAV for chippings	14	12	12	10	10	10
Maximum AAV for aggregate in coated macadam wearing courses	16	16	14	14	12	12

It is not enough merely to provide an adequate macrotexture, however, since if the surfaces of the exposed chippings are themselves smooth, the objective of providing enough frictional resistance is unlikely to be satisfied. Chippings which wear to a smooth, highly polished surface will not perform well. The texture of the surface of individual chippings is of a smaller scale than that of the general surface of the road and is known as the microtexture. Microtexture is measured in aggregates by the polished stone value (PSV) test, whose object is to give a relative measure of the extent to which different types of roadstone will polish under the action of traffic. The test, which is described in BS 812,[15] consists of subjecting samples of stone to an accelerated polishing test in a machine which simulates road conditions, and then measuring the skid resistance value of the polished stone by means of a suitable friction tester. The result of this friction test on the polished stone is reported as the laboratory PSV of the aggregate.

Clearly, the higher the PSV of a material the better is it likely to perform in helping to stop vehicles skidding. Table 2.5 indicates typical minimum requirements for PSV for various locations. Note that naturally occurring aggregates with a PSV in excess of 60–65 are rarely found and that where values in excess of this level are rquired the engineer must turn to the use of synthetic aggregates such as calcined bauxite.

Table 2.5 Recommended polished stone values[14]

Site type	*Traffic cv/lane/day*	*Minimum PSV*
Approaches to traffic signals, pedestrian crossings on major or fast roads	<250	60
	250–1000	65
	1000–1750	70
	>1750	75
Major priority junctions; roundabouts; sharp bends on fast roads; long gradients steeper than 5%	<1750	60
	1750–2500	65
	2500–3250	70
	>3250	75
Elsewhere on roads carrying more than 250 cv/lane/day	<1750	55
	1750–4000	60
	>4000	65
Elsewhere		45

Stone types which are often capable of satisfying the composite requirements of AAV and PSV include basalt, gritstone, porphyry and quartzite, although even within these categories attention should be paid to the quality which may be expected from any particular quarry.

2.8 Bituminous materials

By this term is meant materials which consist of a mixture of bituminous binder and one or more types of aggregate.

There are two very general groups into which bituminous materials are classified – asphalts and coated macadams.

Coated macadams consist of graded stone, coated and to a greater or lesser extent stabilised by a film of binder. They are developments of the unbound surfaces of the last century (see Chapter 6), to which tar or bitumen was added in a haphazard fashion. Modern coated macadams are rather more sophisticated and are specified in the UK by BS 4987.[16]

Coated macadams include materials which by virtue of the grading of the aggregates are dense or open, medium or close graded, and which by virtue of the nominal maximum size of the aggregate and mix proportions are suitable for use in the roadbase, base course or wearing course of a pavement.

The other class of bituminous material in general highway use is rolled asphalt which consists of graded particles of aggregate set in a matrix of binder. Here again the aggregate grading and mix proportions may be varied according to the application and again rolled asphalt is widely used both for surfacing and as a roadbase material.

The different materials have widely differing characteristics and not all materials are suitable for any particular application. Strength is generally a function of material density and so we may expect rolled asphalts and dense coated macadams to be stronger than, for instance, open graded bitumen macadam. However, the stronger materials tend to have higher binder contents and are therefore more expensive. The choice of material should be made carefully.

The design of bituminous pavement elements is discussed in Chapter 4.

2.9 Cement bound materials

There are three general groups of materials used in the pavement which consist of mixtures of aggregates and cement.

Pavement quality (PQ) concrete, for use in rigid pavements to provide the major structural element, is generally specified in the terms of BS 5328.[17] Wet lean concrete, for use as the roadbase in flexible pavements or as the subbase, usually in rigid pavements, is also specified in the UK in the terms of BS 5328.

Dry lean concrete, cement bound granular material, and soil cement differ from normal concrete in that the water content and workability at the time of laying are much less than those of PQ or wet lean concrete. Different specification requirements are therefore appropriate for these three materials, which are considered later.

2.9.1 Pavement quality concrete and wet lean concrete

We require concrete to have a number of properties. Strength is clearly important, but also of concern is the workability of the mix, which will affect the ease with which it can be placed and compacted; and, where the concrete will form the running surface, the durability of the material and its resistance to weathering. Colour can sometimes be of interest. By varying the mix proportions (of coarse aggregate, fine aggregate, cement and water) and the nature of the constituent materials it is possible to control these properties of the finished material to a considerable extent.

The construction process is such that total control cannot be achieved and it is likely that the actual properties of various samples of concrete will be scattered on either side of those required. This is particularly so in the case of the strength of the finished concrete, where variations can arise from differences in the quality of the materials used, in the mix proportions and in the sampling and testing process. The variation in concrete strengths has been found to be similar to the normal distribution as shown in Fig. 2.11.

In Fig. 2.11, the area beneath the curve represents the total number of test results. The number of results which may be expected to be less than any

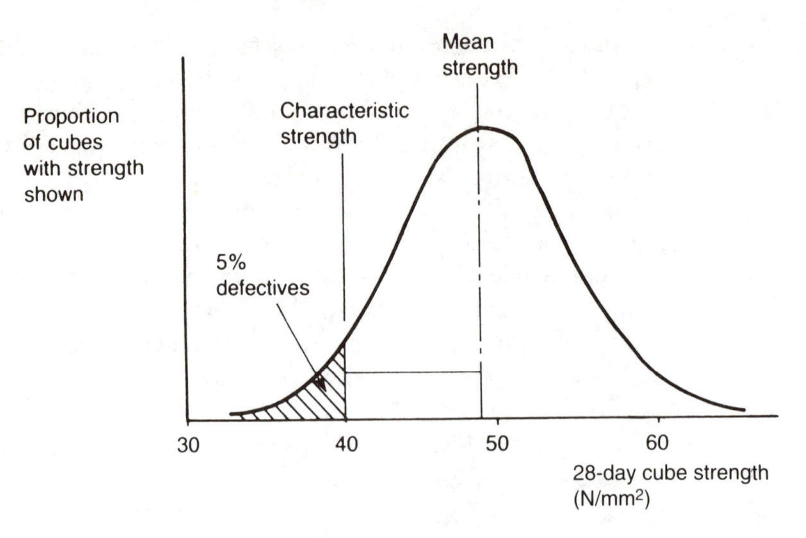

Figure 2.11 Usual variation in concrete strengths for a particular mix

particular compressive strength is proportional to the area under the curve and to the left of the vertical line passing through that strength. Concrete is specified in terms of its characteristic strength, which is that below which only five per cent of the sample is expected to fall. The relationship between this failure rate and the mean may be expressed as a function of the standard deviation of the test results:

$$f_m = f_c + ks$$

in which f_m is the target mean strength (N/mm^2)
f_c is the specified characteristic strength (N/mm^2)
k is a constant, equal to 1.64 if 5 per cent defectives are acceptable
s is the standard deviation, typically between 4 and 6 N/mm^2

Thus whereas the specification will refer to concrete in terms of its characteristic strength, the mean strength used in designs such as those set out in Chapter 5 will be somewhat higher.

The strength of concrete increases with time, as the chemicals in the cement gradually react. Strength measurements for specification purposes are taken 28 days after the concrete has been mixed and placed. This is inconvenient for quality control purposes on site, where the 7-day strength is generally used to obtain an initial indication of concrete strength. Tests should be carried out to determine the relationship between the 7-day strength and the 28-day strength: typically the former will be about 75 per cent of the latter.

Pavement quality control is required in the UK to have a characteristic 28-day strength of 40 N/mm^2, and thus in the terms of the BS is known as C40 concrete. This is a high value in general civil engineering terms, chosen to ensure that the material is durable. Wet lean concrete has a lower strength.

In specifying a mix, certain information should be given to the supplier in order that the design requirements should be met. This information should include:

- *Type of mix* For highways work the general arrangement is for the mix to be designed by the supplier, to meet the specified strength and other requirements;[1]

- *Permitted types of concrete* Ordinary Portland cement (OPC) is generally acceptable, as are many other hydraulic binders. Rapid hardening and high alumina cements are not, unless allowance is made for their high early strength and the amount of cement is increased to provide the necessary durability;

- *Permitted types of aggregate* Aggregates should be such as will enable the pavement to resist wear and should be consistent in their thermal coefficient of expansion with other aggregates in other layers of the same structural element. Very hard aggregates make difficult the cutting of slots for joints and the like. Natural aggregates as previously discussed are generally suitable, as is crushed blast-furnace slag;

- *Grade of the concrete*;

- *Minimum cement content* Cement content affects concrete durability and is measured in terms of mass of cement per unit volume of concrete. For C40 concrete the cement content should be at least 320 kg of OPC per cubic metre; for lean concrete cement content may be specified in terms of the ratio of aggregate to cement, for example 14:1 for C20 and C15 mixed and 18:1 for C10 and C7.5;

- *Workability* The compacting factor test, specified in BS 1881: part 103[18] is appropriate. Target values for compacting factor will depend on the mixes and materials and will typically be of the order of 0.8, with a slight reduction for the lower layer of two-layer construction;

- *Air content* Up to 5 per cent may be accepted;

- *Water/cement ratio* This has been discussed previously in this chapter. Typical maxima are 0.5 for C40 concrete and 0.6 for wet lean concrete;

- *Rate of sampling* Typically one sample per 600 square metres or at least six per day. Samples for workability and air content may be taken at a higher rate.

2.9.2 Cement-bound material – dry lean, cement bound granular and soil-cement

These materials, although cement-bound, are handled and compacted as are unbound granular materials, except that the roadbase materials may be laid by a paving machine similar to those used for bituminous materials.

Typical strength requirements, expressed in terms of the average 7-day compressive strength of a group of five cubes are 4.5 N/mm^2 for soil cement, 7.0 N/mm^2 for cement bound granular material and 10.0 N/mm^2 for dry lean concrete. Recommendations[1] for aggregate grading for the three materials are shown in Table 2.6.

Table 2.6 Aggregate grading for cement bound materials

BS Sieve size	*Soil cement*	*Cement bound granular material*	*Dry lean concrete*
50 mm	100	100	100
37.5 mm	95–100	95–100	95–100
20 mm	45–100	45–100	45–80
10 mm	35–100	35–100	
5 mm	25–100	25–100	25–50
2.36 mm		15–90	
600 μm	8–100	8–65	8–30
300 μm	5–100	5–40	
150 μm			0–8
75 μm	0	0–10	0

2.10 Geotextiles

The word 'geotextile' was coined awkwardly from the Greek and Latin to describe woven materials used in conjunction with soils as part of a construction project. Today a large range of fabrics and sheet materials are available. These differ widely in appearance, manufacture and function.

With one or two exceptions, geotextiles are made from synthetic polymers developed since the 1940s. These are initially extruded to form single threads (filaments) which may then be treated in various ways to give the finished product. These filaments may be of a single material (monofilament) or, where the properties required of the fabric are not consistent with any one polymer, the filament may consist of a core of the base polymer with one or more sheaths of secondary material (the whole being known as a heterofilament). The filament may be twisted with others to form yarn, or it may be used alone; either being woven to form a fabric of the familiar kind or formed into felt-like materials by an appropriate bonding technique – thermal (melt-bonding), chemical (resin-bonding) or mechanical (needle-punching).

The effect of the different approaches to the manufacture of geotextiles is that two groups of material are available: woven and unwoven. Woven materials have relatively regular fibre spacing and high resistance to deformation in the directions of the warp and the weft but not in other directions. Unwoven materials have their constituent fibres arranged randomly within the plane of the fabric and consequently the fibre spacing is less predictable than the woven equivalent, and the ability to resist stress is omnidirectional. The actual spacing of the fibres and the yield strength of the material will of course additionally depend on the nature, size and disposition of the individual fibres and it is for this reason that in describing or specifying geotextiles mention is often made of their weight per unit area.

An analogous but distinct type of membrane is available in the geogrid. Geogrids are formed by extruding the polymer not in the form of a filament but rather as a sheet; holes are then punched in the sheet which is subsequently drawn both parallel and perpendicular to its major axis, the effect being to produce a rectangular grid which, due to the molecular orientation which takes place during the drawing process, exhibits a high resistance to further deformation.

2.10.1 Uses of geotextiles and geogrids

Manufacturers have not been slow to identify many uses. Highway engineers find four general applications as part of the pavement or associated works.

As a separation layer both geotextiles and geogrids may be used to prevent two distinct adjacent groups of particles from mixing. An example of this application is found in temporary roads where the membrane prevents the loss of granular material into soft underlying clay, thus prolonging the useful life of the pavement. This application is discussed in Chapter 6. As a filter a geotextile can allow the passage of water across the plane of the fabric while preventing the movement of solid particles. However, this filtering action will only be successful if the openings in the geotextile are correctly proportioned in relation to the size of the particles to be retained. For this reason geogrids are not appropriate for use as filters and unwoven geotextiles may not be so reliable as the woven alternative with its more controlled filament spacing.

In order that the effectiveness of a geotextile as a filter may be predicted, two characteristics of the fabric should be considered. Firstly, it should be sufficiently permeable to allow water to flow at all; commonly one seeks to ensure that Darcy's coefficient of impermeability for the fabric should be at least five times greater than that for the soil. Secondly, as mentioned above, the range of sizes of the openings in the geotextile is clearly important. What is less clear is the exact mechanism whereby water can continue to flow despite the retention of soil particles at or close to the geotextile. It is thought that a bridging zone of coarse particles accumulates immediately upstream of the geotextile and that this bridging zone itself tends to retain finer particles, preventing these from entering and clogging the geotextile (see Fig. 2.12).

In practice there will be a variation in opening sizes in the geotextile; this variation may be relatively small, as in the case of a woven material, or larger in unwoven materials. In either case it is necessary to relate this distribution of opening sizes to the particle size distribution of the soil to be retained. The opening sizes in the fabric are assessed by passing glass ballotini through it in a manner akin to sieving, the result being expressed in terms of the O_{90} pore opening – that which corresponds to the diameter of the glass sphere fraction of which 10 per cent by weight passes through the fabric. This quantity is related to the soil grading by the use of a reconstructed grading curve for the soil, which is drawn as in Fig. 2.13 tangentially to the midpoint of the grading curve of the soil. It has been found that for effective filtration

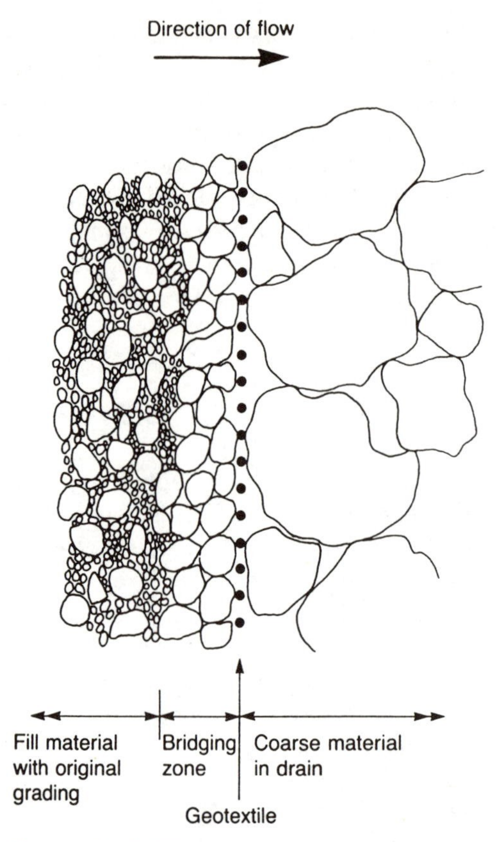

Figure 2.12 Filtering action of a geotextile

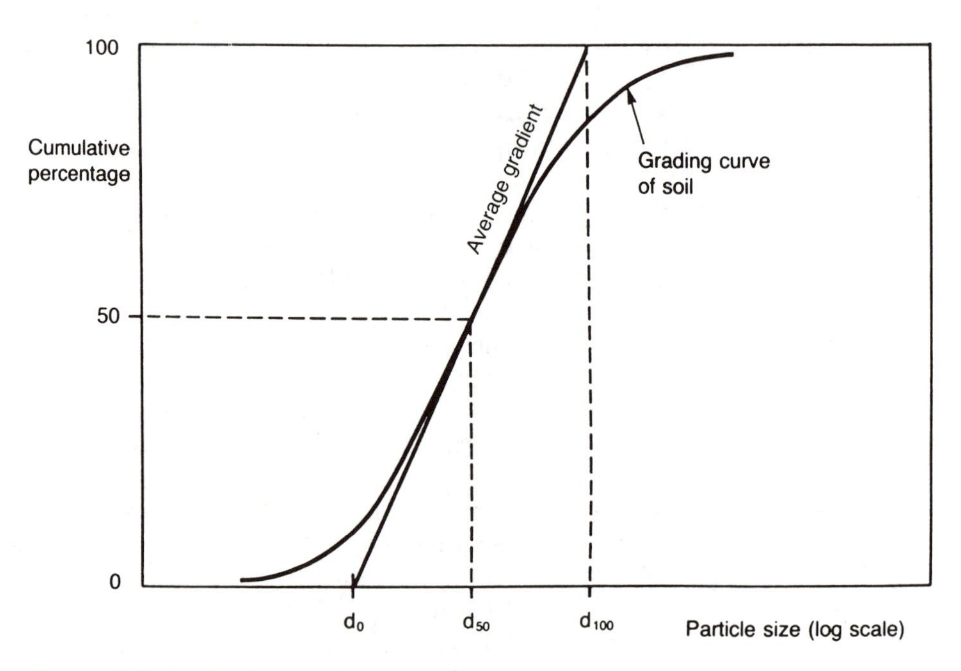

Figure 2.13 Derivation of d_0, d_{50} and d_{100}

to take place, O_{90} should not be greater than d_{90} and that O_{90} should not be greater than ten times d_{50}. Typical specification requirements are that O_{90} should lie between 100 and 300 μm.

As a drainage medium, a geotextile may be used to provide a permeable plane in a body of soil along which subsurface water will flow. Where the intention is to intercept water flowing horizontally, the plane of the geotextile will be vertical and associated with a pipe at the bottom of the fabric (the composite arrangement being known as a fin drain, see Fig. 7.14); where the intention is to intercept water draining downward through the soil, the plane of the geotextile will be horizontal (an arrangement known as a drainage blanket; see Fig. 7.14). In either case, the geotextile is contrived to provide a free-draining layer of material whose thickness might be of the order of 5 mm, sandwiched between two membranes, at least one of which will be a geotextile selected to act as a filter. Sometimes there will be a filter on each side of the 'sandwich' whereas in other cases where the possibility of water passing completely through the drain is inconvenient, one side of the 'sandwich' may be impermeable. This latter arrangement is often used adjacent to retaining walls.

Fin drains provide a cheap alternative to french drains and may be used to provide cut-off drains at either side of a pavement. This is discussed in Chapter 7.

As reinforcement a geotextile or geogrid may be used to enhance the tensile strength of an otherwise weak material. Geogrids are most suitable for this application since the large openings in the material allow interlock and cohesion to develop between the material on either side of the grid. This is discussed in the context of reinforced asphalt in Chapter 4.

2.10.2 General requirements of geotextiles

In the past the strength testing of geotextiles has been carried out in various manners derived from the textile industry. A widely accepted test is the wide strip test, in which samples 200 mm wide and 100 mm long are tested in tension at a standard rate of strain of 10 per cent per minute, the load at 5 per cent strain to be not less than 2.5 kN in the case of material to be used as a separation layer.

Other properties arising out of the nature of the material include the sensitivity of all polymers to ultraviolet light, which can cause serious weakening (some products are protected with carbon black). Chemically contaminated ground or very acid soils with a pH less than about 3 can also cause problems. No life form visible to the naked eye will willingly eat polymers, but in some circumstances microbes present in organic soil will. The first documented use of polymer based geotextiles was in the mid-1960s and there is of course no absolute proof of longevity for a greater period.

Where a stress approaching the breaking stress of the material is applied to a polymer fabric, creep will occur. The exact behaviour depends of course on the particular material; polyester filaments when tested under a constant tensile load of 20 per cent of the breaking load, stabilise at about 3 per cent elongation in a relatively short period – a few days – whereas polypropylenes exhibit much less stability, elongations of over 20 per cent being achieved under similar loading applied during many months. There are clear implications for the choice of materials to be used for reinforcement, and the stress levels that can be supported in the long term.

Revision questions

1 Describe:
(**a**) dry-bound macadam;
(**b**) water-bound macadam;
(**c**) type 1 granular material;
(**d**) type 2 granular material.
How is the stability of these materials ensured in service in the pavement?

2 Briefly explain the difference between cutback bitumens and pen grade bitumens. Give an example of the use of each.

A wide variety of bituminous materials may be used in pavement construction. Give an example of a suitable appliction for each of the following:

- rolled asphalt
- fine cold asphalt
- mastic asphalt
- dense macadam
- open graded macadam
- bitumen emulsion.

Explain your choices.

3 Geotextiles may be used as filters, as separation layers and as tensile reinforcement. Give an example of each application in the context of highway engineering, and describe the benefits associated with each of these examples.

4 Explain the importance of particle size distribution for granular materials used in road construction, with special reference to:
(**a**) subbase materials;
(**b**) bituminous macadam;
(**c**) concrete block paving.
The three curves in Fig. 2.14 indicate results obtained from three materials tested

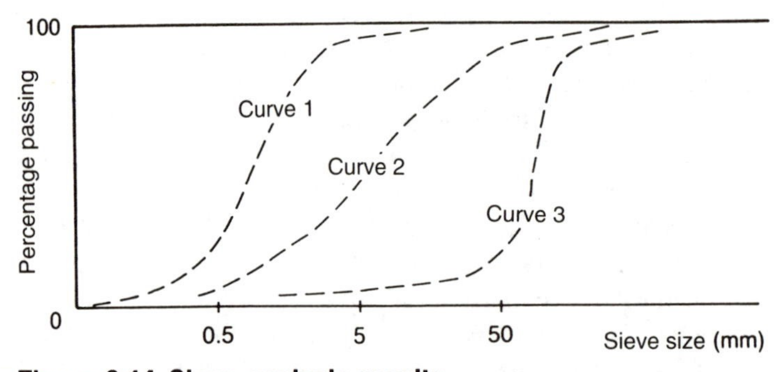

Figure 2.14 Sieve analysis results

in accordance with BS 1377. Assuming all other requirements are met, which of these materials would you accept for use on site in:
(**d**) the subbase of a pavement;
(**e**) concrete block paving;
(**f**) fill to land drains?
Give you reason in each case.

References and further reading

1 'Specification for Highway Works', Department of Transport, HMSO, 1991.
2 Parsons, A. W. and Boden, J. B., 'The moisture condition test and its potential applications in earthworks', Transport Research Laboratory, Supplementary Report 522, 1979.
3 Powell, W.D., Potter, J. F., Mayhew, H. C. and Nunn, M. E., 'The structural design of bituminous roads', Transport Research Laboratory, Laboratory Report 1132, 1984.
4 Croney, D. and Jacobs, J. C., 'The frost susceptibility of soils and road materials', Transport Research Laboratory, Report LR90, 1967.
5 BS 812, 'Testing Aggregates' part 124, 1989, 'Method for the determination of frost heave', BSI.
6 Neville, A. M., 'Properties of concrete', Pitman, 1981.
7 Van Der Poel, C. 'A general system describing the visco-elastic properties of bitumens and its relation to routine test data', Journ. App. Chem. 4, 1954, pp. 221–36.
8 Bitumen test data chart, Shell Research NV, Koninklijke/Shell Laboratorium, Amsterdam. Unpublished.
9 Brown, S. F., 'An introduction to the analytical design of bituminous pavements', 3rd edition, University of Nottingham, 1983.
10 Szatkowski, W. S., 'Resistance to cracking of rubberised asphalt: full-scale experiment on trunk road A6 in Leicestershire', Transport Research Laboratory, Report LR 308, 1070.
11 Lea, F.M., 'The chemistry of cement and concrete', Arnold, 1970.
12 Higgins, D., 'Cement replacements keep salt out', *Highways*, September 1986.
13 'Specification for air-cooled blast furnace slag aggregate for use in construction', BS 1047, 1983.
14 'Specification requirements for aggregate properties and texture depth for bituminous surfacings for new roads', Department of Transport Technical Memorandum H16/76, 1976.
15 BS 812, 'Testing aggregates', part 114, 1989, Method for determination of the polished-stone value, BSI.
16 BS 4987, 'Coated macadams for roads and other paved areas', BSI, 1988.
17 BS 5328, 'Concrete', BSI 1990, 1991, or an equivalent national standard.
18 'Methods of testing concrete: method for determination of compacting factor', BS 1881, Part 103, 1983.

3. Pavement foundation

3.1 General principles

This chapter discusses the lower layers of the pavement, including the subgrade, capping and subbase.

The permanent design requirements of these elements have been stated in Chapter 1 as being to provide a stable platform upon which the construction of the upper layers of the pavement may take place; and in the case of the capping and subbase to regulate the surface of the subgrade and to insulate the subgrade against the action of cold weather. There may also be a requirement to use the completed pavement foundation as a haul route during the construction period. The design of such unpaved temporary roads is discussed in Chapter 6.

Since the stresses induced in the pavement foundation during the construction of the upper layers are to a large extent independent of the design life of the pavement and of the cumulative traffic for which it has been designed, there is no need to attempt to design this lower layer of the pavement in accord with these factors. Rather, the major determinant of foundation thickness is the nature of the subgrade. Experience shows that for subgrades with CBRs greater than about 5 per cent a subbase depth of about 225 mm is required;[1] Brown[2] suggests 200 mm.

However, if the subgrade CBR is found to be less than 5 per cent this provision of subbase cannot be expected to be adequate. An inadequate subbase may exhibit during construction such signs of distress as deformation (either at the surface during its compaction or in the bound layer above the subbase during compaction) or, in extreme cases, heave in the subgrade where this is in soft clay and where a large compactive effort is applied to the subbase. This heave is caused not by the action of frost but by the applied loading. In order to avoid these problems an increased thickness of foundation is provided.

Since each of the modes of failure mentioned above is a manifestation of excessive subgrade strain, the designer should address the issue of what constitutes the maximum acceptable level of strain in order to ensure that this level is not exceeded. The amount of strain that will develop in any particular case will be a function of the magnitude of the applied loads and the way in which the soil responds to those loads. This response is related to the elastic modulus of the soil, represented by E, which according to Black and Lister[3] is related to the plasticity index of a cohesive soil by the equation

$$E = 70 - PI \qquad [3.1]$$

where PI represents the plasticity index, expressed as a percentage, and E is measured in MPa. There is thus an implicit relationship between the strain at the formation level and the CBR of the soil.

The influence of loads applied directly to the top of the pavement foundation during construction is less easy to assess but in situations where the subbase is not used as a haul route or similar and the design philosophy outlined in Chapter 6 does not apply, the adequate assumption is made that the influence of construction traffic is not a significant variable. From site to site, stresses at the formation will vary, owing to differences in the thickness of the pavement foundation layers above. Stresses at the formation due to loads applied at the surface of the subbase will of course bear an inverse relationship to the thickness of the subbase and capping.

Unsurprisingly, the conclusion of all this is that weak soils with low CBRs (and low elastic moduli) demand thicker pavement foundations than do stronger soils with higher CBRs.

The practice is often adopted of dividing the pavement foundation layers into two: the subbase towards the top of the foundation consisting of high quality well graded materials such as granular material type 1, or bound material, and lower down in the foundation, in the interests of economy, using cheaper materials of lower quality. A common requirement for this lower, capping, layer is that it should be of adequate grading to allow compaction – no very large particles should be included – and that it should itself be unlikely to fail in service as a result of internal vertical deformation. A minimum CBR of 15 per cent is often required to meet this proviso. The material should in addition be frost resistant.

These considerations give rise to the standard designs illustrated in Table 3.1.[1]

Table 3.1 Recommended thickness of subbase and capping

Subgrade CBR (%)	*Capping thickness (mm)*	*Subbase thickness (mm)*
0–2	600	150
2–5	350	150
5–15 (pavements with granular subbases)	0	225
5–15 (concrete pavements with lean concrete subbases)	150	150
Over 15 (all pavements)	0	150

3.2 Sources of capping material

Since the specification for the material is less stringent than is the case for subbase materials, cappings may be of significantly cheaper materials than those used in more highly stressed upper layers and the engineer may therefore exercise some ingenuity in seeking appropriate sources of material. A source similar to that employed for the subbase may not be attractive because of the relatively high cost in comparison with lower grade stone.

Material types which have been successfully used in cappings include those which occur naturally in the vicinity of the works, naturally occurring materials found remote from the site, crushed concrete, graded hardcore, and salvaged road materials from either on or off site. Recycled materials may usually be expected to have a CBR well in excess of the commonly required 15 per cent although it should be noted that recycled and pulverised bituminous materials – obtained for instance from the cold planing of mature surfaces – may exhibit surprisingly low CBRs when tested. This is because the behaviour of the material is influenced in the test by the presence of the binder which coats the aggregate and which deforms in a viscous manner when subjected to the shearing stresses set up in the test. Nevertheless, such materials used in cappings have performed satisfactorily.

We shall see (Chapter 9) that the vertical strain caused in a pavement by an applied load is related to the probable service life of that pavement. An effect of using a free-draining, relatively coarse-grained material is that any water present in the material is readily dispersed when external loads are applied, and as a consequence it is found that vertical strains in the material are small. If however the proportion of fines in the material is allowed to rise then this free-draining condition is lost to an extent and we find that the permanent vertical deformation in the layer due to an applied load increases. From this is deduced the notion that contamination of the subbase or capping is a bad thing in that it tends to reduce pavement life.

Contamination of the layer immediately above the formation may be caused by pumping of clay particles up into the structure of the pavement resulting from repeated loading of the system in the presence of water. This has caused the premature failure of a number of pavements, including motorways designed before the problem was properly recognised. The usual approach is to provide a filter at the formation to prevent the upward movement of clay particles. Traditionally, sand filters have been provided but a more convenient and widely used alternative is offered by the use of a geotextile at the formation to act as a filter during the life of the pavement, and also to enhance the ability of the pavement foundation to act as a haul route during the construction phase. Bell[4] has found that the clay contamination in a granular material above a filter bears an approximate linear relationship to the 95th percentile pore size of the filter – the finer the filter, the less the contamination. Geotextiles are discussed in Chapter 2, where it is indicated that the finest pore sizes are achieved in woven materials rather than the more randomly arranged unwoven types.

The provision of a suitable geotextile at the formation is thus likely to increase the life of a pavement built on cohesive soil and is recommended. In addition there are benefits to be obtained during construction from the use of a geotextile in this way since geotextiles can make a significant contribution to the strength of an unbound road.

It is often the case that at sites where the formation is of a plastic clay which demands a capping there is no suitable local material available to form this part of the pavement. The engineer then has the choice of excavating the appropriate volume of unsuitable material and arranging for its disposal, probably off site, and importing a similar quantity of capping material – an expensive option – or, in some circumstances, modifying the properties of the native soil so that it becomes suitable for use in the capping.

3.3. The modification of subgrade properties

The benefits to be obtained from these techniques are those of cost – the amount of material to be handled can be reduced – and sometimes of reduced nuisance due to construction traffic, arising out of the reduced volume of material to be carried to and from site.

The methods which can be used to achieve these benefits include the permanent stabilisation of the soil by mixing with cement, lime or less commonly bitumen; or by the use of various mechanical methods.

3.3.1 Cement stabilisation

By this term is meant the process of mixing cement with a native soil or imported aggregate so as to produce a cement bound material (CBM) whose strength is much greater than that of the original unbound material, although

Table 3.2 Indicative guide to suitability of cement stabilisation of different soils

		Suitability			
Construction method	*Mixing plant*	*Clean granular*	*Granular with up to 15% <75 μm*	*Soil with up to 35% <75 μm and PI < 10*	*Silty clays*
Mix-in-plant	Free-fall concrete mixers	No	No	No	No
	Pan or paddle type concrete mixers	Yes	Possible	No	No
	Pugmill stabilisation mixers	Yes	Yes	Possible	No
Mix-in-place	Agricultural type rotovators	Yes	Yes	No	No
	Purpose-made stabilisation rotovators	Yes	Yes	Yes	Yes

usually much less than that of pavement quality concrete. Unlike PQ concrete, these materials are mixed with a low water content – the expression 'earthdry' often being used to describe this state – and are compacted by rolling.

Not all soils can be successfully stabilised in this way. As a general rule, the further removed a soil is from the 'all-in' aggregate used in the building trade for the *ad hoc* production of concrete, the less likely is cement stabilisation to be successful. Thus for example sandy soils are eminently suitable whereas a soft clay is most unlikely to yield satisfactory results. The mixing method employed is also significant and Table 3.2 offers a tentative guide to combinations of plant and soil likely to result in successful stabilisation. In any specific potential application the soil should be tested; particle size distribution, liquid and plastic limits, soil acidity and sulphate content should all be determined. The physical tests enable an assessment to be made by means of Table 3.2; the chemical tests will point to possible interferences with the hydration of the cement or chemical attack on the hydrated cement. The effect of soil acidity can best be checked by carrying out strength tests on cured samples of the stabilised soil, while soils may be considered capable of being satisfactorily stabilised if the total sulphate content is not greater than 0.25 per cent for cohesive soils or 1 per cent for granular soils.

A typical specification for granular material for cement stabilisation to form capping[6] should be any material other than unburnt colliery spoil and argillaceous rock having the following properties:

Maximum liquid limit	45
Maximum plasticity index	20
Maximum organic matter content	2%
Maximum total sulphate content	1%
Saturation moisture content (chalk)	20%

Grading

Sieve size	% by mass passing
125 mm	100
90 mm	85–100
10 mm	25–100
600 μm	10–100
63 μm	0–10

If tests indicate that cement stabilisation of the soil is likely to be viable, it remains to determine the desirable mix proportions which will result in a successful CBM; that is, the target moisture content and cement content which must be achieved during construction. There are two ways to approach this problem.

The optimum moisture content of the mixture of soil and cement will of course differ from that of the soil alone since the fine-grained cement will demand proportionately more water than will the soil. It is therefore necessary to make an initial estimate of the cement content which will be required and to proceed on that basis, modifying results as necessary later. Moisture/density tests are carried out on the mixed material in a manner similar to that described in Chapter 2 for soils; a full description of the method for soil/cement mixtures is given in BS 1924 'Stabilised materials for civil engineering purposes'.

Having established the likely optimum moisture content, specimens are prepared and compacted at this moisture content with the expected necessary proportion of cement present and strength tests carried out on these specimens. If the strength requirements are economically met then the specimen mix proportions are accepted as the design proportions; if not then the process is repeated until a satisfactory result is obtained.

There are various ways of defining what constitutes an appropriate strength requirement. It is often convenient to express specification requirements in terms of the crushing strength of cubes of the material prepared in a standard way, for example in accordance with BS 1924. This approach has the advantage that a full evaluation of the materials in terms of the specification may be achieved in the laboratory or testing station at relatively little cost and in standardised conditions. If such an approach is to be adopted, a common requirement is that the strength of the material thus indicated should be of the order of 3.5–4 N/mm^2 after seven days. However, this approach has the drawback that it does not directly relate to the strength requirements of the material as expressed in the pavement design for the capping – a minimum effective CBR of 15 per cent.

The second approach to the design problem is a more pragmatic one particularly suited to the mix-in-place method of construction. Here one ensures that the amount of free water present in the system is enough to facilitate mixing of the material, hydration of the cement and adequate compaction of the CBM. The cement content is initially set at 2 per cent by weight of dry soil, or other value considered suitable, and a test area of CBM is constructed. The CBR of the material in this area is then checked against that specified. The initial cement content may of course be determined by laboratory tests, related either to crushing strength, as before, or to CBR.

3.3.2 Construction of CBM

Two approaches are again possible: mix-in-plant, where the material is mixed away from the point of laying to which it is transported to be laid and compacted; or mix-in-place, where all operations take place at the point of laying. The former is most appropriate where imported granular materials are used; the latter where native soil is to be stabilised.

In-plant mixing may take place on or off site. In either case the most important requirement of the mixing plant is that it should be capable of thoroughly mixing the constituent materials – plants with a positive mixing action are necessary since the material is much less fluid than mixed concrete, for which a simple tumbling action is usually sufficient. Once mixed, the material must be transported to the point of laying by means which will avoid segregation and significant changes in the moisture content. On arrival the material is spread on the prepared supporting layer and compacted. Spreading may be by plant fitted with blades or shovels, which can with care achieve satisfactory results and which can be arranged so as not to traffic the supporting layer. Alternatively, and especially for the larger job, paving machines may be used but it should be noted that these of necessity traffic the lower layer of material, as must delivery vehicles. In addition some plant operators consider that the use of pavers in conjunction with non-bituminous materials causes unacceptable wear on the conveyors and augers in the machine. The use of a paving machine is nevertheless often attractive because of the relative ease with which a uniform layer of material may be produced.

Plant-mixed CBM will generally be laid in such a way that, for adequate compaction to be achieved at the edges of the material, a restraint of some sort will be necessary – otherwise the unsupported material at the side of the layer will simply move outwards under the action of the roller resulting in poor compaction at the edges. Edge restraints may be either side forms firmly fixed and set at least to the level of the uncompacted material, or they may be provided by spreading the CBM over an area slightly wider than that finally required – an extra width of about 300 mm on each side is usually enough – so that the extra material can itself provide the necessary restraint for the body of the work. After compaction these extra margins are removed and the layer cut back to the required width.

During the layer process, attention should be paid to two further matters. There will be an initial surcharge of material, before compaction has taken place, and this should be allowed for in arrangements for edge restraints and the like. The degree of surcharge will vary according to the material and the degree of initial compaction; in its uncompacted state a typical CBM may require 15–20 per cent surcharge in order that the required final thickness may be achieved, whereas if a paving machine is used there will be an element of intial compaction and the surcharge will be correspondingly reduced. Secondly, in all but the smallest job, there will of necessity be joints in the material. These will be either transverse – 'end of day'–joints, where the work is interrupted, or longitudinal joints between one pass and another of the spreading plant. End of day joints should be arranged across the full width of the work and should be of such a form that a vertical face of fully compacted material is presented to the continuation of work at a later date. This is usually achieved by ramping down the end of the work and then cutting away the excess. The use of transverse forms can result in difficulties in compaction at the end of the work and should be avoided unless special arrangements can be made for compaction plant to run off the work. No expansion or contraction joints are necessary in CBM, but it should be noted that feathered joints can lead to vertical movement due to expansion of the material. Longitudinal joints will be necessary where the spreading operation cannot be to the full width of the pavement and should consist simply of laying fresh material against previously laid uncompacted material and then compacting the two together. Not more than one hour should elapse between the laying of adjacent areas in this way.

CBM should be compacted by similar plant to that required for the compaction of unbound materials, within two hours of laying, until maximum density has been achieved. Vibrating rollers are most widely used; steel tyred deadweight rollers are often avoided since they may sink into or disrupt the uncompacted material. One or two passes of a vibratory roller will cause the material to bed down sufficiently for the full compactive effort to be applied. After compaction the material should be cured by the application of a suitable membrane to prevent moisture loss before hydration of the cement has taken place.

Mixed-in-place CBM has much in common with plant-mixed material but of course differs in that the mixer passes over the site, accepting as it does so a suitable combination of soil, water and cement and mixing these to leave behind a material in a fit state for compaction. *In situ* mixing plant consists essentially of a rotovator, which like the familiar horticultural machine uses rotating tines to break and mix the soil. There the similarity ends, however, since machines used in highway construction are generally more powerful by several orders of magnitude, being capable of stabilising clays and granular materials up to 350 mm deep and of breaking bound material of various types. Agricultural rotatators may be used for thinner layers, up to about 150 mm, in conjunction with suitable soil types.

The construction process consists of preparing the site, adjusting the moisture content of the soil, spreading cement on the soil, and mixing the two together. Compaction and curing then follow as in the case of plant-mixed material.

Site preparation should ensure that the soil is in a proper state for the process. Topsoil should be removed, the soil shaped to the required profiles, and if necessary the moisture adjusted by the addition of water or allowing the soil to dry. In areas of heavy soil where slow working is expected, it may be worthwhile to pulverise the soil before applying the cement in order that the mixing process may not be unnecessarily protracted; this pulverisation may be adequately achieved by rotovating the soil.

After cement has been spread at the required rate, the components are then mixed by passing the rotovator over the work. More than one pass of the machine may be necessary. Care should be taken to ensure that passes of the machine overlap, horizontally and vertically where appropriate, in order that a continuous layer of material is produced. The alternative of a number of smaller and therefore weaker elements adjacent to or superimposed on one another is unlikely to be satisfactory.

Cement stabilised soils have been widely used in cappings and subbases with considerable success. As a capping material cement has most often been used in conjunction with imported relatively low grade granular materials since the soft cohesive soils which demand the use of cappings cannot be used to form a successful CBM. The greatest economy that can be achieved with such soils by the use of stabilisation techniques is made by the use not of cement but of lime.

3.3.3 Lime stabilisation

The effect of lime on clay has been briefly described in Chapter 2; essentially the effect is that much of the available water is lost from the clay, whose particles flocculate to form individual granules with a consequent increase in the CBR of the soil. It is by no means uncommon to find the CBR increased from less than 5 per cent to 30 per cent or more within three days of the addition of 2 per cent by weight of lime.

Again the suitability of the native soil for this method of stabilisation must be determined. Lime stabilisation will only work with soils which contain enough clay for the reaction described above to take place; attempts to use the lime as a binder, in a way analogous to the use of cement, will not be successful. The moisture content of the soil is also important since this will affect the moisture condition value (MCV) of the soil and render compaction of the stabilised material difficult.

A typical specification[6] for cohesive material for stabilisation with lime to form a capping is that any material other than unburnt colliery shale might be used, having the following properties:

Minimum plasticity index	10
Maximum organic matter content	2%
Maximum total sulphate content	1%
MCV immediately before compaction	12 or less
Grading	
Sieve size	% by mass passing
75 mm	100
28 mm	95–100
63 μm	153–100

The construction process is always of the mix-in-place type and bears many similarities to that for cement stabilisation. There are however a number of significant differences.

The most important of these arises out of the violence of the reaction of lime with water. Consequently the material is dangerous and operatives exposed to quicklime can experience severe burns, both externally and internally, and blinding. The importance of care in handling lime, and of ensuring that all those who come into contact with lime wear proper protective clothing including dust masks and protective goggles, cannot be overemphasised. The danger passes as stabilisation continues and subsequent construction work may proceed normally.

Stabilisation consists of preparing the clay to accept the lime – by breaking up the soil where necessary with a suitable pulveriser – spreading the lime at an appropriate uniform rate, mixing the lime and soil by further passes of the pulveriser, and lightly rolling the surface to enable site traffic to pass over it and to seal it against the ingress of further water. During the next few days the reaction between the lime and clay will proceed and the CBR of the mixture will gradually increase. The rate of increase will depend on the nature of the soil, the amount of lime present, and the ambient conditions, but typically after three or four days a CBR of at least 10–20 per cent should have been achieved. The material is then re-pulverised to facilitate compaction until the specified MCV is achieved, and finally compacted and trimmed.

The target rate of spread of the lime should be determined by prior tests, the objective being to achieve a particular CBR. Typically, between 2 and 4 per cent by mass is required – about 5–10 kg per square metre of soil per 150 mm depth to be treated. The lime is usually provided in granules sized between 6 and 20 mm; larger lumps do not promote easy mixing while smaller particles can be blown by the wind which is both wasteful and dangerous.

Once the second stage of the process is complete the surface of the work can be trafficked by construction plant immediately.

Generally the most satisfactory result is obtained by the use of a purpose built rotovator, similar to that described for use in cement stabilisation and in Chapter 9 for pavement recycling, since this plant will achieve the most intimate mixing of lime and clay. However, in some circumstances much may be achieved by more rough and ready means such as working lime into the surface of the soil by passing and repassing with a tracked vehicle. This approach is often satisfactory for the provision of temporary haul routes across areas of soft clay, and similar applications.

Lime stabilisation has been widely used throughout the USA and elsewhere. In the last few years the technique has been re-introduced to the UK and the process is now accepted for use on trunk road works. The economics of the process are such that it becomes a feasible alternative to the removal and replacement of material in areas in excess of about 2500 m^2 in suitable soils.

3.4 The subbase

General requirements of the subbase have been discussed at the beginning of this chapter. Specific details relate to the choice of subbase material.

For flexible pavements a suitably graded granular material is an appropriate choice and should be capable of functioning satisfactorily. Suitable granular soils were described in Chapter 2 and are specified in the UK by the 'Specification for Highway Works' as subbase Type 1 or Type 2. Of these, Type 1 is generally the most suitable because of the high degree of interlock which can be developed between aggregate particles and because of its free-draining nature. However this material is not always the cheapest and in areas where natural sand/gravel mixture are freely available Type 2 is often used for less taxing applications such as car parks and lightly trafficked roads.

For reinforced concrete pavements there is another requirement of the subbase not mentioned hitherto – it must be capable of allowing construction of the concrete slab to proceed satisfactorily. Because of the different techniques used, the requirements of the subbase as a construction platform differ in this case in that a firm base is needed on which reinforcement can be located prior to the placing of concrete, and a surface of low permeability should be presented to the freshly poured concrete in order to minimise grout loss. Although arrangements can be made satisfactorily in these respects with a granular subbase (rendered impermeable at the top with plastic sheeting, reinforcement placed on spacers of sufficient size to avoid penetration of the subbase), an alternative exists. This is to form subbase in lean concrete (see Chapter 4), or very occasionally in the form of a blinding layer of low-grade concrete, suitably jointed. This latter approach is used only in the more informal projects but lean concrete subbases have often been specified for motorway and trunk road schemes. In order that compaction of the lean concrete should be carried out effectively and without damaging the subgrade, the material should be laid to the thicknesses indicated in Table 3.1.

Current requirements[7] for concrete pavements in trunk roads and motorways in the UK are that the subbase should be in wet lean concrete or cement bound granular material.

3.5 Subsurface water

It has been noted in Chapter 2 that the CBR of most soils is influenced to a greater or lesser extent by the amount of water present. Protection of the subgrade from excess moisture is of fundamental importance in good highway design.

Often it is the case that the moisture content of the soil beneath a wide impermeable pavement remains effectively constant; drying by surface evaporation is prevented by the pavement, as is percolation down from the surface to the subgrade. However, in circumstances where lateral movement of subsurface water is likely this will not be the case – there will be an appreciable risk of the subgrade's ability to resist vertical strain being greatly reduced, with possibly disastrous effects. In such circumstances the provision of protective measures to de-water the subgrade should be considered.

Historically, ditches have been provided at both sides of the road to drain the subgrade and prevent subsurface water from adjacent areas seeping under the road; today these are less attractive than formerly because of the need for regular maintenance and the availability of suitable substitutes. These substitutes are discussed in Chapter 7.

3.6 Construction of capping and subbase

The sequence of events in the construction of the pavement foundations is generally of the following nature:

(1) Bulk earthworks to provide a surface about 300 mm above the final formation;
(2) Provision of subdrains, surface water drains and ducts beneath the pavement;
(3) Reduction of earthworks outline to final formation levels;
(4) Preparation of formation;
(5) Placing and compaction of capping (if any);
(6) Placing and compaction of subbase.

We are concerned here with items 4, 5 and 6.

3.6.1 Preparation of formation

In order that the subsequent layers of the pavement can be properly constructed it is important that the formation is in a fit state to receive them. It should be fully compacted, uniform in profile and have a moisture content consistent with the pavement design. Failure to achieve these requirements may result in premature failure of the pavement caused by excessive vertical stain due to either the poor condition of the subgrade or the underprovision of pavement foundation where poor level control at the formation has occurred. Preparation of the formation can consist of regulation of the subgrade surface by careful trimming and rolling with a smooth wheeled roller. In less fortunate circumstances extensive works may be necessary in the removal of soft spots, reducing the water content of the subgrade, and similar measures.

Rain is the great enemy here. An increase in the moisture content can change a satisfactory formation to a highly unsatisfactory one overnight, and waiting for the soil to dry out again can cause serious delay to a contract. The prudent contractor therefore arranges his operations so that the formation is left exposed to the elements for as little time as possible; once a satisfactory formation has been achieved it should be covered with the next layer in the pavement as soon as possible. Cohesive soils in particular are vulnerable to changes in moisture content and difficult to dry out once waterlogged.

3.6.2 Placing and compaction of unbound materials

Here the objective is to achieve full compaction (up to the specified requirement, typically 95 per cent of the maximum dry density) through the full depth of the material.

The amount of compaction achieved at any part of a body of soil will depend on the nature of the soil in question, the intensity of the stress tending to cause compaction, and the number of times that stress is applied. Thus a compliant soil subject to frequent large stresses will tend to be better compacted at the end of the process than will a stiff soil, subjected to one or two applications of a relatively small stress. The construction process will consist of driving an item of plant over the surface of a layer of initially uncompacted material. The variables which the engineer can significantly control are the nature and condition of the material, the thickness of the layer, the intensity of loading caused at the surface by the plant, and the number of times the plant passes over every point in the layer. For compaction to take place there must also be a firm base to resist the compactive effort of the plant; this will be provided by a properly treated formation.

The influence of moisture content on the ease with which a granular soil may be compacted has already been discussed. Typical specification requirements in this respect depend on the amount of cohesive matter likely to be present in the material and are illustrated in Table 3.3.

Apart from moisture content requirements, unbound granular materials of broadly similar gradings respond to compaction in similar ways.

The thickness of the layer of material is of course very significant. If the layer is too thick for the compactive effort expended then the lower parts of

Table 3.3 Typical specification requirements for moisture content of unbound materials at compaction

Material	*Required moisture content*
Granular subbase Type 1	Often none stated
Granular subbase Type 2	Optimum moisture content -2% $+1\%$
Wet mix macadam	Optimum moisture content $\pm 0.5\%$

the layer will be incompletely compacted, and settlement may occur when the pavement is in service. If the layer is too thin it may be deformed by the compaction plant. A suitable range of thicknesses appropriate to normal civil engineering plant lies between 110 mm and 225 mm. Thicknesses greater than 225 mm should be laid and compacted in two or more layers each not less than 110 mm.

A variety of plant is available for use in compaction. The appropriate choice will be determined by the nature and scale of the work. Each type of plant

Table 3.4 Compaction requirements for unbound pavement foundations[6]

		No. of passes for layers not more than the following compacted thickness (mm)		
Type of compaction plant	*Category*	*110*	*150*	*225*
Smooth-wheeled roller	Mass per metre width of roll:			
	2700–5400 kg	16	N	N
	5400 kg+	8	16	N
Pneumatic-tyred roller	Mass per wheel:			
	4000–8000 kg	12	N	N
	8000–12000 kg	10	16	N
	12000 kg+	8	12	N
Vibratory roller	Masses per metre width of vib. roll:			
	700–1300 kg	16	N	N
	1300–1800 kg	6	16	N
	1800–2300 kg	4	6	10
	2300–2900 kg	3	5	9
	2900–3600 kg	3	5	8
	3600–4300 kg	2	4	7
	4300–5000 kg	2	4	6
	5000 kg+	2	3	5
Vibrating plate compactor	Mass per square metre of base plate:			
	1400–1800 kg/m^2	8	N	N
	1800–2100 kg/m^2	5	8	N
	2100 kg/m^2	3	6	10
Vibro-tamper	Mass:			
	50–65 kg	4	8	N
	65–75 kg	3	6	10
	75 kg+	2	4	8
Power rammer	Mass:			
	100–500 kg	5	8	N
	500 kg+	5	8	12

N = not suitable

has its own unqiue performance characteristics. Smooth wheeled deadweight rollers combine a high contact pressure with the ability to produce a smooth uniform surface in bound materials. Pneumatic tyred rollers have much lower contact pressures and may tend to disturb the surfaces of unbound materials. Vibratory rollers superficially resemble deadweight rollers but differ in that within one or more of the drums is an eccentric weight rotating much faster than does the drum and (ideally) independent of the motion of the roller. The effect of this eccentric weight is to increase greatly the efficiency of the roller, owing in part to the dynamic loading to which the material beneath is subjected and in part to oscillations set up in the material if the applied vibration approximates to the natural resonant frequency of the material. A frequency of about 25 Hz is appropriate for crushed rock; 33 Hz for a cohesive sand/gravel/clay mixture. Vibrating plate compactors and power rammers tend to be used for small areas of work, inaccessible or uneconomic for larger plant.

The appropriate number of passes of any of these items of plant will depend on the thickness of the layer to be compacted. Table 3.4 indicates current UK requirements.

Revision questions

1 List and briefly explain the factors that you would need to consider when designing and specifying a capping layer and a subbase for a new road. To what extent do the design and specification of the unbound layers in the pavement influence the design and specification of the bound layers?

2 How would you treat the formation
(**a**) in areas of fill?
(**b**) in areas of excavation with a high water table?
State the objectives you are seeking, why these objectives are important, and how you would ensure that they are achieved.

3 What factors will determine the viability of soil stabilisation as an alternative to the use of imported material in the pavement foundation? Identify those factors which depend on the nature of the materials present, and those which depend on other site factors.

References and further reading

1 Powell, W. D., Potter, J. F., Mayhew, H. C. and Nunn, M. E., 'The structural design of bituminous roads', Transport Research Laboratory, Laboratory Report 1132, 1984.
2 Brown, S. F., 'An introduction to the analytical design of bituminous pavements', 3rd edition, University of Nottingham, 1983.
3 Black, W. P. M. and Lister, N. W., 'The strength of clay fill subgrades: its prediction in relation to road performance', Transport Research Laboratory, Laboratory Report 889, 1979.
4 Bell, A. L., McCullough, L. M. and Snaith, M. S., 'An Experimental Investigation of Subbase Protection Using Geotextiles', *Proc. 2nd International Conf. on Geotextiles*, Las Vegas, IFAI Vol. 2, 1982.

5 Kennedy, J., 'Cement-bound materials for sub-bases and roadbases', Cement and Concrete Association, 1983.
6 'Specification for Highway Works', Department of Transport, HMSO, 1986.
7 Department of Transport Departmental Standard HD 14/87 'Structural Design of New Road Pavements', Department of Transport, 1987.

4. Flexible bases and surfacing

4.1 Introduction

A flexible pavement is one capable of retaining its structural integrity when small vertical movements take place at the surface. In the case of public roads in the UK this usually implies a bituminous surfacing which will itself be supported by one or more base layers above the unbound pavement foundation. In many respects the bituminous layers can be considered as constituting one element of the pavement. This simplification is a convenient one when the overall strength of the pavement is to be considered, although later in this chapter we shall consider in detail the individual elements of the bound layer.

A wide variety of bituminous materials are available, many of which are to a greater or lesser extent suitable for use in various ways in the bound layer. The essential problem which the engineer should solve is the selection of a suitable material for each pavement element, and the determination of a suitable thickness of that element. This is best done in an awareness of what is required of the bound layer.

4.2 Characteristics of a successful bound layer

As in any engineering exercise, cost is important here – both in terms of the initial construction, and of the likely subsequent maintenance needs. In order to satisfy various different demands the engineer often provides a composite layer of material.

The purpose of the pavement is to allow traffic to pass safely and comfortably at reasonable speed. From the user's point of view it is therefore important that the upper layers of the pavement should present a surface which has suitable characteristics such as surface regularity and skidding resistance to the degree appropriate to the function of the road. These requirements will vary in detail from one application to another – thus an interurban trunk road with good alignment might demand a different set of physical surface characteristics from those which are appropriate for a poorly aligned urban route, even though the volume of traffic carried might be the same.

To achieve this objective, it is equally important that the structure of the pavement, particularly the upper, bound, layers should be strong enough to support traffic loads and transmit these to the layers beneath in such a way as to avoid premature failure of the pavement. In Chapter 1 the concept of the finite life of pavements was discussed, and once the engineer has selected an appropriate design life all reasonable steps should be taken to ensure that this target is approximately reached; the thickness and strength of the bound layers are major factors here.

While a pavement may satisfy both these criteria at the beginning of its life, deterioration will start as soon as it is built. Clearly the designer should bear this in mind and should ensure that pavements are sufficiently durable to last for an appropriate period without deformation, cracking or loss of structure texture. If the subgrade is also to perform as expected it is important to ensure that adequate protection is provided against surface water; the surface of the pavement should provide a waterproof surface as a protection for the underlying layers.

General design practice is to seek to satisfy the strength requirements of the pavement wholly within the roadbase, the surfacing not being considered to make any significant contribution to the strength of the pavement; and to provide a suitable wearing course at the surface whose prime function is to ensure that the surface is of appropriate texture and regularity for the intended traffic.

4.3 Roadbase

The roadbase makes the major contribution to the strength of the pavement.

The design problem which the engineer should overcome is to ensure that appropriate materials are used, and that the roadbase is neither too thick nor too thin. If too great a thickness is provided then the cost of the project is unnecessarily high; if the roadbase is too thin it will either fail within itself, usually by cracking, or it will fail to provide adequate protection to the unbound layers beneath. Rutting will then occur initially at the formation, to be transmitted through the pavement structure to the surface. Ways of approaching this design problem are characterised by two general methods.

Historically, design methods have been sought by studying the performance of pavements in service and deducing design charts from the variety of results obtained. An example of such an approach was the work by the then Road Research Laboratory in the study of the behaviour of lengths of experimental pavement on the A1 trunk road at Alconbury, Cambridgeshire, during the 1960s. This and similar work led to the publication in 1970 of the third edition of Road Note 29[1] which represented the zenith of the empirical approach to the problem. This third edition and its predecessor formed the basis of conventional pavement design philosophy in the UK during the period 1965–1984, during which years most of the UK motorway network was built for the first time. Experience showed this empirical approach to be lacking – largely because it was developed from a narrow data base.

The fieldwork which led to this empirical approach was of necessity limited by the resources of the researchers to the relatively detailed consideration of pavements made of a few different combinations of a few different materials, built and used in the temperate conditions of the UK, and trafficked by up to about 10 million standard axles (msa). Such a philosophy does not make for innovations in design or in the selection of materials for the pavement; nor does it necessarily provide competent designs for pavements expected to carry well over 100 msa as is often the case today.

Figure 4.1 illustrates the variations in construction which would have arisen from the use of standard UK pavement design methods current during the lives of pavements likely to be found in service today. This diagram shows an increase in pavement thickness of about 30 per cent between the 1965 design method and that of 1984. While some of the difference between the various designs can be accounted for by the differences in VDF over the period considered (see equation [1.1]), and by changes in the definition of 'design life', at least some of the change which has occurred can be attributed to an increased knowledge of the way in which the various materials behave.

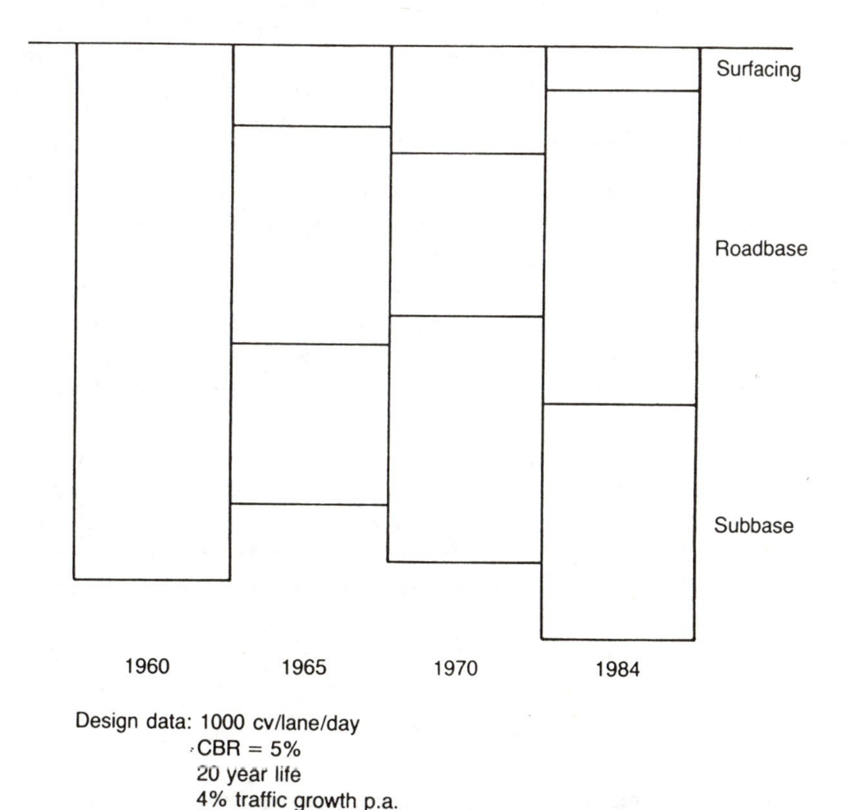

Figure 4.1 Variations in standard UK pavement design methods

It is always comforting to believe that the current state of knowledge is adequate and that nothing worthwhile remains to be added. Unfortunately this is hardly ever the case; certainly not in the field of pavement design.

However, in other areas of engineering where less reliance is placed upon empiricism design methods tend to last longer and, it might be inferred, tend to model actual events more closely. This is achieved by a theoretical analysis of, for example, the stresses set up in each element of a structure by a system of applied loads and a comparison of these with the stresses which the material is known to be capable of supporting – the design being modified as necessary until a sound and economic arrangement is achieved. It is in order to bring about such a state of affairs in highway engineering that the problem of the analytical design of pavements has been addressed in various quarters.

If such an analytical design method could be produced in a workable and reliable form suitable for use in all construction situations then it would be a very powerful tool for the designer; the way would be clear for a much wider use of a far greater variety of materials than is encouraged by the current conventional methods. For example, the use of polymer modified binders can produce asphalts which are more stable and resilient than those which contain bitumen as the only binder. Since, however, the former do not feature to any great extent in studies of pavement behaviour in service, they are not considered in any authoritative empirical design methods and are therefore little used, thus ensuring that the opportunities to study their behaviour in service remain infrequent. This circular argument could be broken in this and similar cases if an acceptable way to predict adequately the performance of such a pavement were available.

The development of an analytical design method for flexible pavements is a complex problem because the materials to be considered are neither homogenous nor Hookian in their behaviour; there is no simple elastic response to loading, nor can failure be readily predicted, arising as it often does out of material fatigue rather than simple tensile or compressive breakdown. Current analytical design methods are characterised by their corresponding complexity and, more significantly from the point of view of the engineer, sensitivity to a number of variables. These are either beyond his control or difficult or impossible to quantify in the context of the construction process.

Nevertheless, consideration of at least one analytical design method is of value since the mechanism will be illustrated by which failures eventually occur in flexible pavements and the material characteristics which tend to resist these will be discussed. Current pavement design methods[7,12] are derived from an amalgam of the more reliable elements of empiricism and current analytical design practice.

4.4 An analytical design method for bituminous pavements

One of the most widely known methods is that developed at the University of Nottingham.[2]

The method is intended to predict the thicknesses of the bound and unbound layers necessary to prevent failure of the pavement during a service life equivalent to the passage of a given number of standard axles over the surface.

Modes of failure considered are fatigue failure in the bituminous bound layer, resulting in cracks starting to develop at the bottom of the layer and propagating to the surface – a failure mode which results from the development of excessive horizontal strains at the bottom of the bound layer (see Fig. 4.2(a)); and permanent vertical deformation of the subgrade which reflects through the flexible pavement to cause rutting at the surface – a failure mode associated with excessive vertical strain at the formation (see Fig. 4.2(b)). The method consists essentially of calculating the maximum strain caused by the application of a single standard axle which can be accepted at the formation (vertical strain) and at the bottom of the bound layer (horizontal strain), and ensuring that sufficient thicknesses of material are provided to prevent these strains being exceeded.

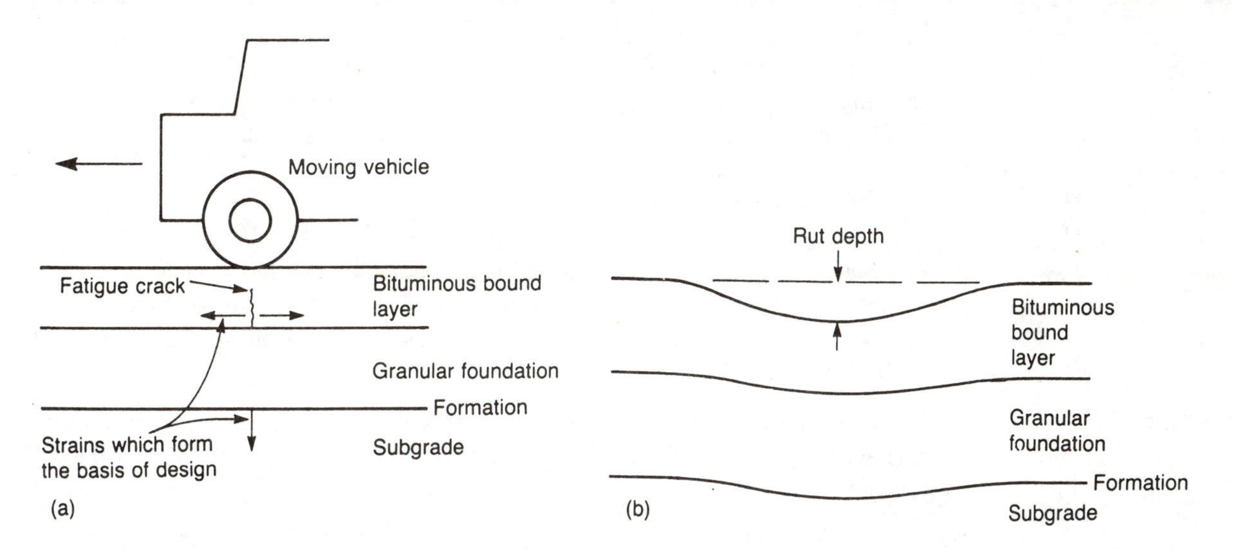

Figure 4.2 Failure modes in bituminous pavements (a) Fatigue crack and design strains (b) Permanent deformation

The assumption is made that the various layers of material – roadbase and surfacing – which make up the bound layer act as one, and that there is no significant variation in strength between one component part of this composite layer and another. This assumption is often valid, but we shall see later that in the case of the more sophisticated designs developed for the most heavily trafficked pavements, materials are used in such a way that some caution is appropriate in the use of the method.

A further assumption made in this method is that the pavement foundation is equivalent to an unbound layer 200 mm thick. Variations in this quantity are not likely to affect the required thickness of the bound layer significantly since the strength of the pavement foundation is much less than that of the bound layer.

4.4.1 Calculation of maximum allowable strain in bound layer

Here we are concerned to ensure that the material used has fatigue characteristics such that premature failure will not occur. Fatigue failure is caused by repeated applications of loads none of which is large enough to cause failure when applied alone. The ability of a bituminous material to resist tensile forces is provided by the binder and not by frictional forces between the aggregate. Since we are concerned with the repeated application of a standard (40 kN) load which will itself cause some finite degree of strain it is clear that the number of load applications which the pavement can tolerate in this respect will be inversely related to the degree of strain caused by each single application; that is

$$N_{\mathrm{f}} = f(1/\mathrm{strain}) \qquad [4.1]$$

in which N_{f} represents the number of load applications to failure and the function $f(\ )$ has been found to depend on the amount of binder present in the mix, and its viscosity. Cooper and Pell[3] carried out extensive tests on bituminous materials and developed the empirical relationship

$$\log(\mathrm{strain}) = \frac{14.391 \log V_{\mathrm{b}} + 24.21 \log \mathrm{SP}_{\mathrm{i}} - 40.7 - \log N}{5.131 \log V_{\mathrm{b}} + 8.63 \log \mathrm{SP}_{\mathrm{i}} - 15.8} \qquad [4.2]$$

in which V_{b} represents the percentage of binder by volume
SP_{i} the softening point of the binder in its initial state – before laying
N the number of load applications corresponding to the given level of strain

A successful design in terms of fatigue failure will be which one does not allow this maximum allowable strain caused by the passage of one 40 kN wheel load to be exceeded.

4.4.2 Calculation of maximum allowable subgrade strain

Here too actuality has so far defied successful analysis. Brown[2] quotes an empirical relationship which relates the vertical strain caused by the formation by the application of a single 40 kN wheel load at the surface to the number of load applications associated with the onset of critical conditions (N):

$$\text{Max. formation strain} = \frac{21600}{N^{0.28}} \qquad [4.3]$$

Once again, the design objective is to ensure that this degree of strain is not exceeded.

4.4.3 Estimation of required bound layer thickness

Having determined the maximum strains that can be tolerated at the bottom of the bound layer and at the formation, it remains to provide material of sufficient strength and thickness to ensure that these strains are not exceeded. By making various assumptions the number of variables in this complex problem can adequately be reduced to three: the modulus of elasticity of the subgrade (which may be approximated to ten times the CBR when this is expressed as a percentage); the thickness of the bound layer; and its stiffness. Estimation of CBR has been discussed, and the thickness of the bound layer is the unknown to be determined. It remains to estimate the stiffness of the mix – that is, the ratio of stress to strain appropriate under the service conditions.

The stiffness of a bituminous mix depends on the stiffness of the binder and the volumetric proportions of the mix – provided that the binder is itself sufficiently stiff to have a major effect. If the binder stiffness is less than about 5 MPa then the situation becomes more complicated. Above this level the mix proportions may be adequately expressed by the term VMA (voids in mixed aggregate), which represents the proportion which the volume of binder plus the volume of air voids bears to the volume of the mix as a whole, expressed as a percentage.

The stiffness of a mix may be estimated from[4]

$$\text{Mix stiffness} = (\text{Binder stiffness}) \frac{(1 + 257.5 - 2.5\ \text{VMA})^n}{n(\text{VMA} - 3)} \qquad [4.4]$$

in which $n = 0.83 \log \dfrac{(4 \times 10^4)}{(\text{Binder stiffness})}$ for stiffnesses in MPa

or the stiffness may be obtained from direct measurement.

To pursue the predictive model it is therefore necessary to determine the likely VMA of the material and the stiffness of the binder to be used. VMA is a function of the degree of compaction achieved in the field and can only be determined from experience gained in field tests or in the use of similar material elsewhere.

Binder stiffness, as has been noted in Chapter 2, is a function of the temperature of the material and of the time of application of a load. The nature of the relationship is illustrated in Fig. 4.3 and is such that the stiffness bears an inverse relationship to time of loading and to temperature; the sensitivity of the binder in these respects is found to be indicated by the penetration index. The relationship has been expressed[5] thus:

$$\text{Binder stiffness} = 1.157 \times 10^{-7} t^{-0.368} 2.718^{-\text{PI}_r} (SP_r - T)^5 \qquad [4.5]$$

in which t represents the time of loading – valid for t between 0.01 and 0.1 seconds
PI_r the penetration index of samples of the binder recovered from use in the field (valid from -1 to $+1$)
SP_r the corresponding softening point
T the temperature – valid for $SP_r - T = 20$–$60\ °\text{C}$

and the stiffness is expressed in MPa.

Evaluation of the three variables on the right hand side of equation [4.5] merits some consideration since the method is sensitive to variations here.

The time of application of the load will depend on vehicle speeds, and on the depth of the bound layer. At the bottom of a thick layer the effect of a passing load at the surface will be felt longer than would be the case in otherwise

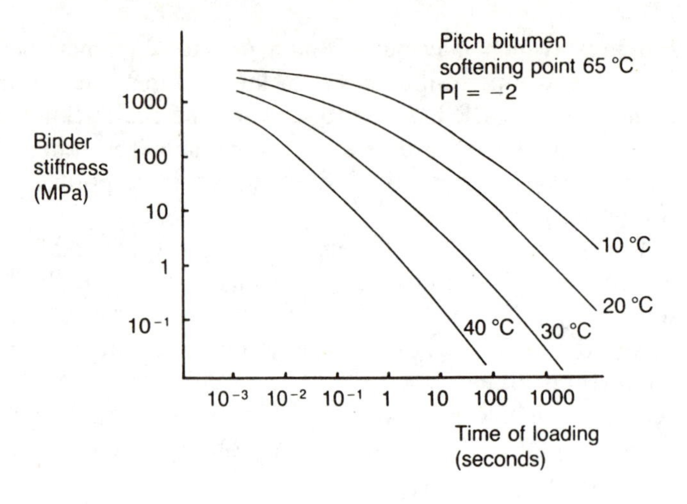

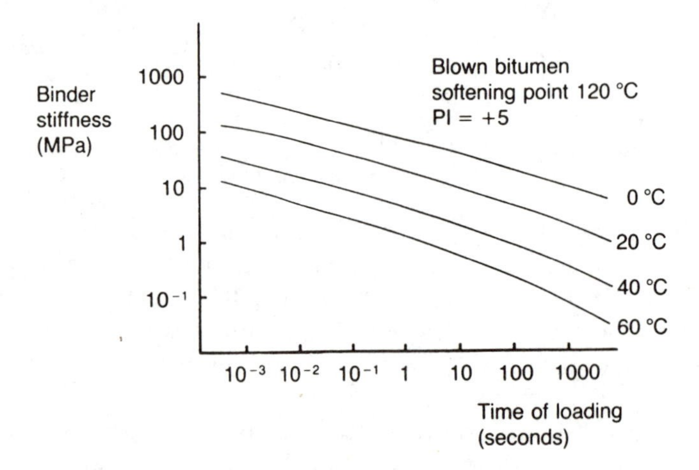

Figure 4.3 Variations in binder stiffness with temperature and penetration index

similar conditions if the layer were thinner. It is possible to estimate the thickness of the layer reliably, if only by iteration. Vehicle speeds are harder to predict. In the absence of an established predictive model, the fiftieth percentile vehicle speed is used.

Brown[2] quotes an empirical relationship

$$t = 1/V \text{ seconds} \qquad [4.6]$$

where V is the vehicle speed in km/hr. This relationship is generally adequate for bound layers between 100 and 350 mm thick.

The assessment of pavement temperature by any means other than direct measurement is complex. This complexity arises out of the number of meteorological variables which come into play, and the thermodynamic behaviour of the pavement/subgrade system. A method has been derived[6] which is capable of a high order of accuracy but, in view of the uncertainty as to how temperature variations relate to times of peak traffic flow (of commercial vehicles), the simplified approach of taking the mean annual air temperature and adding 3 °C to allow for the thermal characteristics of the material, is often adopted.

Softening point and penetration index data on recovered samples of binder are often not available. These may be estimated from penetration tests carried out on unlaid samples of the binder. If this initial penetration is denoted by P_i then the 'recovered' penetration P_r may be estimated from

$$P_r = 0.65 P_i \quad [4.7]$$

and the recovered softening point SP_r may be obtained from

$$SP_r = 98.4 - 26.4 \log P_r \quad [4.8]$$

The penetration index may then be obtained from Fig. 2.9, or by other means.

Equations [4.5] and [4.4] may then be used to predict binder stiffness and thus mix stiffness.

By means of design charts such as that shown in Fig. 4.4 the thickness of bound layer required to resist failure during the design life as a result of subgrade strain, and the thickness required to resist failure due to horizontal strain in the bound layer, may be determined. The greater of these two thicknesses should be provided. Note that Fig. 4.4 represents one of a family of such charts, corresponding to different subgrade moduli of elasticity.

The subgrade modulus of elasticity may be estimated from the following approximation, previously quoted as equation [3.1]:

$$E = 70 - PI$$

where E represents the subgrade modulus, measured in MPa, and PI the plasticity index of the soil, expressed as a percentage.

An illustrative design example may clarify this.

It is required to design a pavement for the following service conditions:

Cumulative traffic	50 msa
Average speed of commercial vehicles	75 km/hr
Subgrade – sandy clay, *PI*	20%
Bound layer – to be rolled asphalt	
Binder	50 pen RB (residual bitumen)
VMA	18%
V_v	5%

Stage 1

Average temperature of bituminous layer = 8 + 3 = 11 °C

Stage 2

Time of application of load

$t = 1/V$ seconds (equation [4.6])

$= 1/75 = 0.013$ seconds

Stage 3

Binder properties

$P_i = 50$ (given)

$P_r = 0.65 P_i$ (equation [4.7]) $= 32$

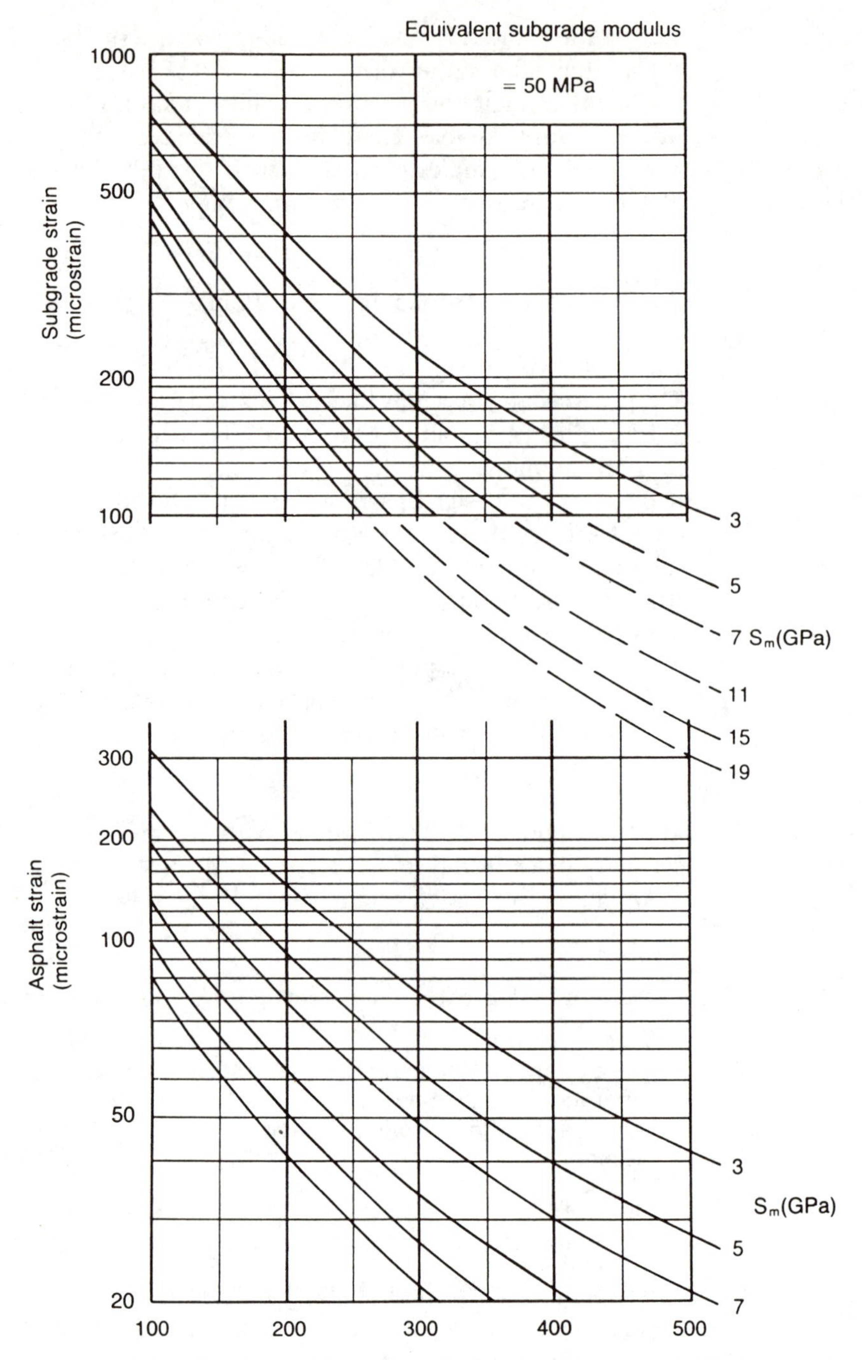

Figure 4.4 Critical strains as functions of stiffness and thickness of asphalt (subgrade modulus of elasticity = 50 MPa)

$SP_r = 98.4 - 26.4 \log P_r$ (equation [4.8])

$= 56.5$

$PI_r = -0.6$ approx (using Fig. 2.9)

Stage 4
From equation [4.5],

$$\text{Binder stiffness} = 1.157 \times 10^{-7} \times 0.013^{-0.368} \times 2.718^{-0.6} \times (56.5 - 11)^5$$
$$= 203 \text{ MPa}$$

Stage 5

In equation [4.4], $n = 0.83 \log \left(\dfrac{4 \times 10^4}{203} \right)$

$$= 1.90$$

$$\text{and mix stiffness} = \frac{203(1 + 257.5 - (2.5 \times 18))^{1.9}}{1.9(18 - 3)}$$
$$= 11.7 \text{ GPa}$$

Stage 6
The proportion of binder present in the mix, by volume, is such that

$$V_b = \text{VMA} - V_v = 13\%$$

Stage 7
From equation [4.2],
Maximum allowable bound layer strain is such that

$$\log(\text{strain}) = \frac{14.391 \log 13 + 24.21 \log 52 - 40.7 - \log(50 \times 10^6)}{5.131 \log 13 + 8.63 \log 52 - 15.8}$$
$$= 1.946$$

Hence maximum allowable bound layer strain = 88 microstrain.

Stage 8
From Fig. 4.4

Limiting condition in respect of bound layer strain = 150 mm thickness

Stage 9
Maximum allowable subgrade strain: from equation [4.3]

$$\text{Max. allowable strain} = \frac{21600}{(50 \times 10^6)^{0.28}}$$
$$= 151 \text{ microstrain in the subgrade}$$

Stage 10
From Fig. 4.4

Limiting condition in respect of subgrade strain = 240 mm thickness

Stage 11
The final design is the greater of the two limiting conditions for the bound layer thickness, in conjunction with a foundation of 200 mm, that is

240 mm thickness of rolled asphalt

200 mm thickness of granular subbase material.

In circumstances where unproven combinations of materials and service conditions will arise, analytical design methods such as this will provide the best means of estimating appropriate pavement proportions. Where tried and tested materials are to be used in commonplace circumstances, empiricism has more to offer and, since the number of variables in the design process has effectively been reduced, there is little point in adopting the more complex, analytical, approach. For example, the pavement design for a car park or any urban road where static traffic may be expected is beyond the scope of equation [4.5] and while it is possible to overcome this difficulty by detailed analysis such an approach is often discarded in favour of a simpler method.

4.5 A standard design method

A current UK design method seeks to take the best of both the empirical and the analytical approaches and to use these to model pavement behaviour, interpreting and extrapolating from the broadened data base now available in the light of analytical research.

The TRRL has[7] provided a family of design charts for pavements built on an adequate foundation (225 mm subbase on a subgrade of 5 per cent CBR, or the equivalent). These are based on the use of materials which comply with the UK Department of Transport Specification for Highway Works and on the ambient conditions during the life of the pavement being similar to those found in the UK or other temperate areas. They also depend on a definition of pavement failure as the latest time when the application of an overlay could be expected to make the best use of the original quality of the pavement in extending its life (see Chapter 9) – this latter corresponding to rut depth in the wheel path in excess of 10 mm, or when cracking is observed. These design charts are reproduced in Fig. 4.5 and represent designs which are considered to have an 85 per cent probability of reaching the required life.

In Fig. 4.5 it should be noted that not all materials are suggested to be suitable for pavements expected to carry more than 20 msa. In the case of lean concrete a standard 'long life' design is offered; the behaviour of lean concrete in these circumstances is complex and not fully understood. Wholly bituminous pavements designed for more than 80 msa are not included in Fig. 4.5 since at this level of traffic a more complex pavement type than the relatively simple system of a homogenous roadbase with a thinner layer of bituminous surfacing becomes appropriate.

At these high traffic flows the consequences of deep-seated failure of the pavement often include severe disruption and delay to traffic while the failed pavement is removed and replaced. The designer may wish to avoid this in which case the design chart shown in Fig. 4.6 should be used.

Field tests have repeatedly shown that in many respects there is little to choose between the performance of dense bitumen macadam as a roadbase and that of rolled asphalt. The macadam is therefore generally used, because its lower binder content makes it cheaper than rolled asphalt. However, the fatigue strength of the cheaper material is inferior to that of rolled asphalt, as might be expected from the analytical work previously outlined. If, therefore, a layer of rolled asphalt is placed at the position in the pavement where a high fatigue strength is most required – at the bottom of the bound layer – then the life of the pavement will be extended and its behaviour modified so

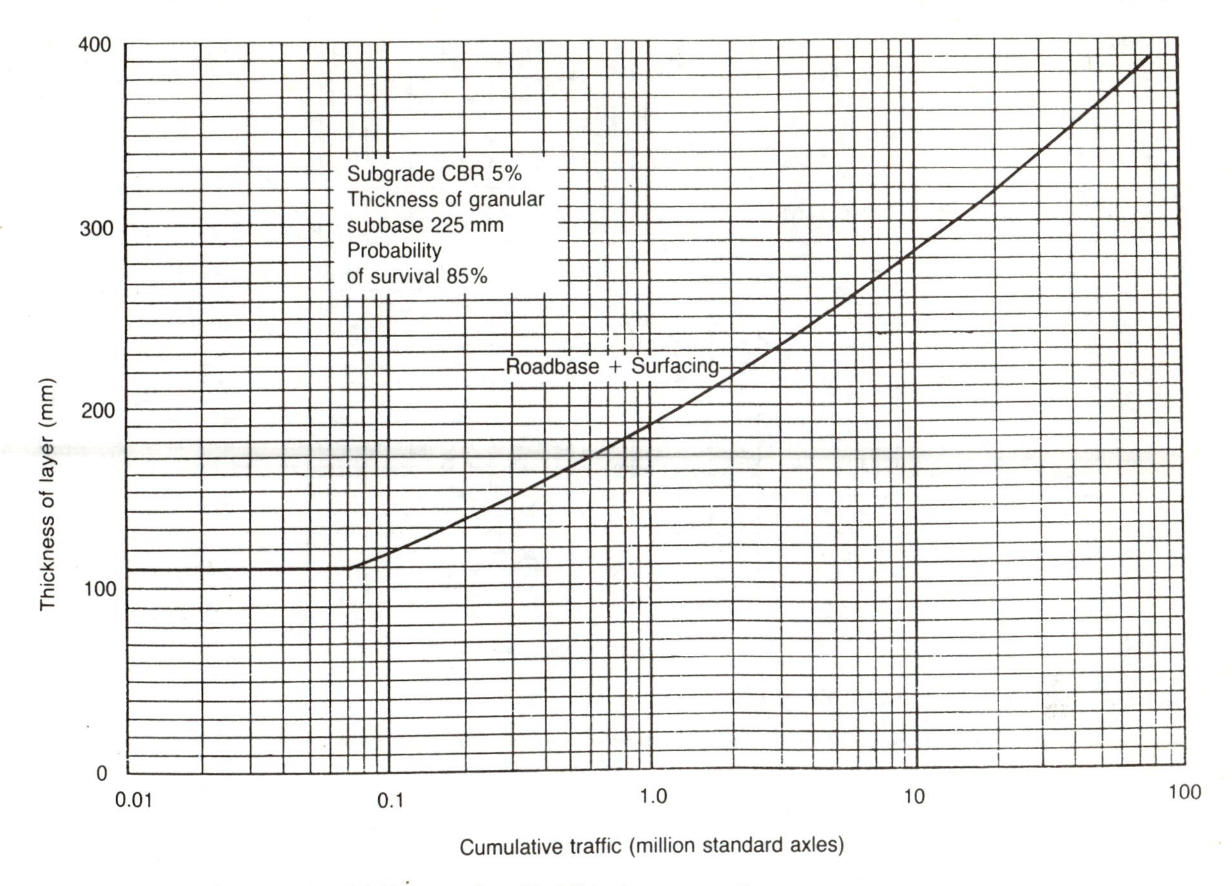

Figure 4.5 Design curves (a) For roads with bituminous roadbase

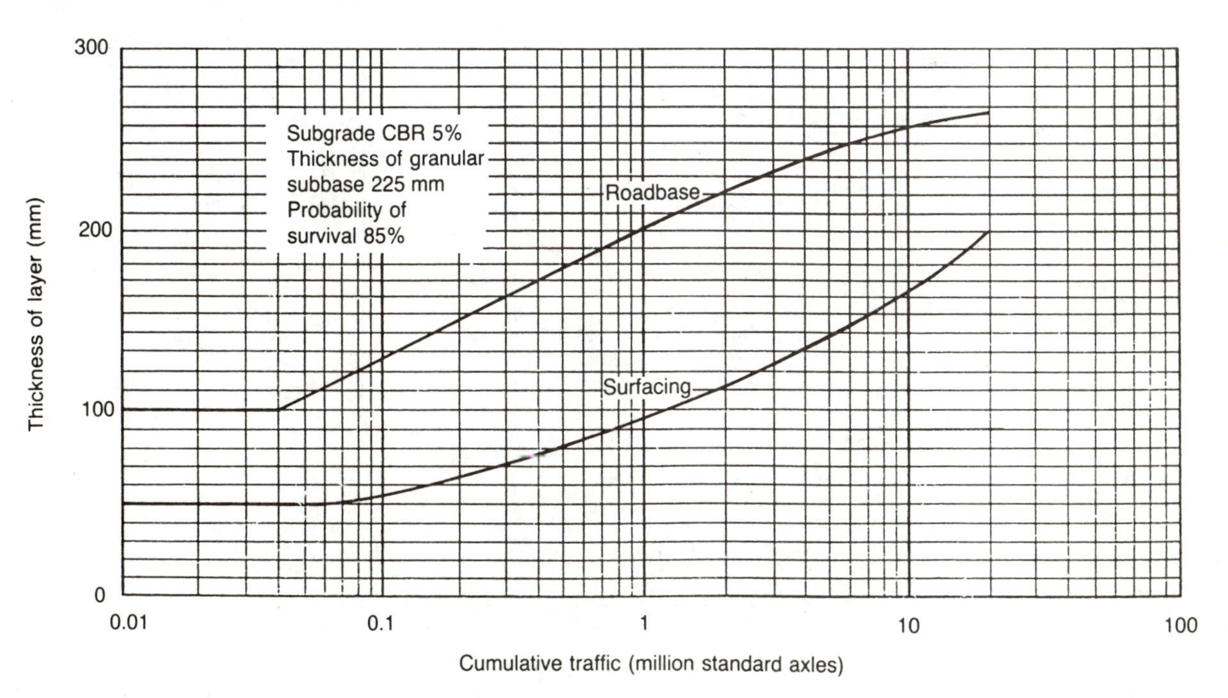

(b) For roads with wet mix roadbase

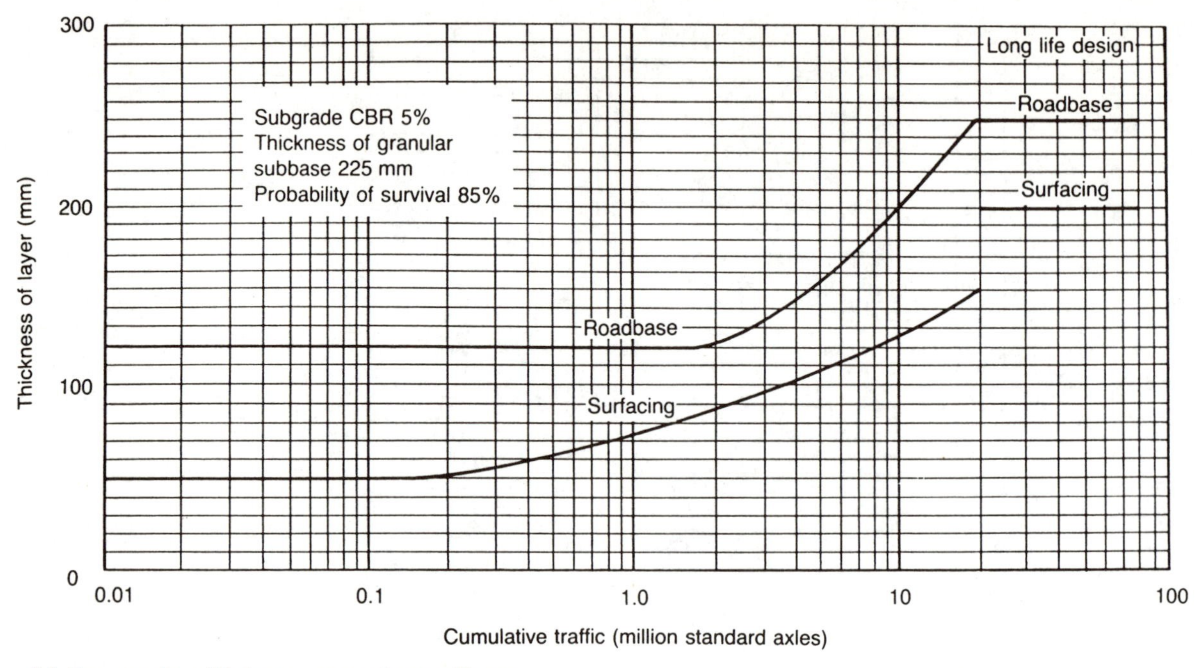

(c) For roads with lean concrete roadbase

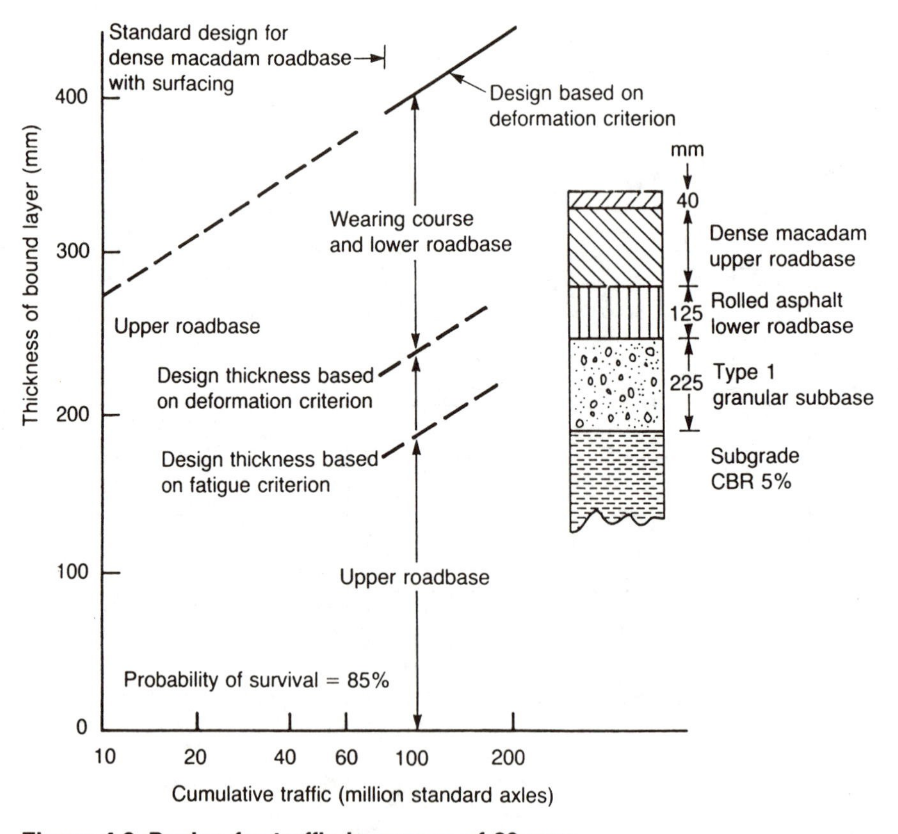

Figure 4.6 Design for traffic in excess of 80 msa

that failure when it occurs is likely to be a result of surface deformation rather than of deep-seated fatigue cracking. Repairs will consequently be confined to the upper region of the pavement, and will involve less cost and traffic disruption than would otherwise be the case.

In Fig. 4.6 are shown three curves. The uppermost indicates the total thickness of the bound layer required to achieve deformation rather than fatigue failure. The two lower curves indicate the necessary thicknesses of dense macadam upper roadbase consistent with ultimate failure by deformation (middle curve) or by fatigue (lower curve). The design uses a standard lower roadbase of 125 mm of rolled asphalt and a 40 mm wearing course, again of rolled asphalt.

The design charts presented in Figs 4.5 and 4.6 are based on the use of appropriate types of materials generally referred to in the charts. If designs produced by these means are to be successful the engineer should ensure that the materials used are indeed appropriate to their purpose in terms of their general type and detailed specification.

4.6 The selection and design of flexible materials

In many cases the engineer will seek to choose suitable materials for the various elements in the pavement by the application of only two criteria: performance and cost. Table 4.1 indicates suitable flexible materials for various applications, deduced from the performance criterion.

It is not always the case that such positive guidance can be offered in terms of the cost of the various options. The practice has therefore developed of designers preparing alternative designs using various acceptable combinations of materials so that tenderers may present the most competitive price consistent with an adequate design.

In some cases there will be additional constraints on the choice of materials – particularly in respect of the wearing course. Matters of concern here often include colour and surface texture.

4.7 The specification of bituminous materials

Asphalts and macadams consist of mixtures of graded stone and binder. Their performance depends on the successful choice of constituent material types, the correct proportioning of those materials, and proper construction. The general properties of the constituent materials were considered in Chapter 2; it remains to be discussed how these are best used in conjunction with one another.

4.7.1 Coated macadams

These are materials in which graded aggregate is coated with a bituminous binder – generally bitumen but tar or a mixture of the two are also used – in

Table 4.1 Flexible pavements – selection of materials

Pavement element	*Traffic (cumulative msa)*	*Suitable materials*
Wearing course	2.5+	Rolled asphalt at least 40 mm thick Previous macadam¶
	0.5–2.5	Rolled asphalt Dense wearing course macadam Close graded macadam Medium graded macadam* Open graded macadam* Minimum thickness 20 mm
	0–0.5	Rolled asphalt Any bituminous macadam*
Basecourse	2.5+	Rolled asphalt Dense basecourse macadam
	0.5–2.5	As above, plus single course macadam
	0–0.5	As above, plus any other coated macadam
Roadbase	Over 80	Composite bituminous – see Fig. 4.6
	20–80	Rolled asphalt Dense roadbase macadam Lean concrete
	0–20	As above, plus Cement bound material Wet mix

¶See section 4.7.1.3
*Seal surface against the ingress of water with grit or surface dressing

which aggregate interlock makes the major contribution to the strength of the material. Coated macadam in which the binder is bitumen is known as bitumen macadam; where tar is used the composite material is tarmacadam.

Coated macadam is classified according to the nominal size of the aggregate, its grading, and the intended use of the material. Open graded, medium graded, close graded and dense materials respectively contain progressively increasing proportions of fines and provide increasingly dense and stable materials. Fine graded macadam is sometimes referred to as 'fine cold asphalt', misleadingly as it is a macadam. Pervious wearing course materials are also available which, by contrast, are almost devoid of fines and are used to prevent spray being thrown up by traffic in wet conditions.

Dense macadam has a carefully graded aggregate and has strength characteristics comparable with, although not equal to, those of rolled asphalt.

Standard mixes are available in three groups – roadbase, base course and wearing course. The significant variables are the nature of the aggregate, and the type of binder.

4.7.1.1 Aggregates

The nominal maximum size of the aggregate should be determined from a consideration of the required thickness of the material when laid, and the surface texture which is to be obtained. Choices which are commonly available include:

For roadbases
- 40 to 28 mm dense road macadam

For base courses
- 40 or 20 mm open graded base course macadam
- 40 mm single course macadam
- 40, 28 or 20 mm dense base course macadam

For wearing courses
- 14 or 10 mm open graded wearing course macadam
- 14 or 10 mm close graded wearing course macadam
- 6 mm dense wearing course macadam
- 6 mm medium graded macadam
- 3 mm fine graded macadam (fine cold asphalt)
- 20 or 10 mm pervious wearing course macadam

Clearly the aggregate size should not exceed the required thickness. In practice the nominal size should be considerably less than the required depth – ideally about half the depth – in order that some degree of interlock should develop between the aggregate particles. A common recommendation is that the thickness of the laid material should never be less than 1.5 times the nominal aggregate size, although this clearly does not apply in the case of fine cold asphalt.

The surface texture of a road is generally required to be coarse in order that skidding resistance will be enhanced. Clearly therefore when a fine-textured material such as fine cold asphalt is used in the wearing course special measures need to be taken to achieve this. Chippings of a coarser stone are thus applied to the surface while this is still soft, and rolled in to provide the necessary texture. Conversely, where a surface is likely to be used by slow moving vehicles and large numbers of pedestrians – such as in mixed use areas – the designer might seek a smaller sized aggregate in order to provide a more suitable surface for pedestrian use. Fine-textured materials are usually more expensive than those of a coarse texture.

As an illustration, let us suppose that the design for a certain residential road is such that 60 mm of surfacing is required. Alternatives include:

- 60 mm of single course macadam, sealed with surface dressing or bituminous grit – the quickest to lay and often the cheapest;

- 40 mm of 20 mm nominal size open graded base course macadam plus 20 mm of 10 mm nominal size open graded wearing course plus seal in bituminous grit – two-course construction means that the final surface can be provided after site traffic has used the road;
- 40 mm of 20 mm base material as above, plus 20 mm of fine cold asphalt. Two-course construction as above, plus fine textured surface.

Different circumstances suggest which is to be preferred.

The aggregate grading for coated macadams tend to the 'continuous' form – similar to the general type for Types 1 and 2 granular materials – in order that aggregate interlock may be promoted. Suitable aggregate types include crushed rock, slag or gravel.

4.7.1.2 Binders

The nature of the binder will of course affect the material. On one hand it is a good thing to keep viscosity of the binder as low as possible in order that the material may be handled and compacted at low ambient temperatures; on the other hand a high viscosity is required after laying, in order that the surface is not damaged by the passage of traffic before it has cured (that is, before the volatile fractions have left the system of a cutback binder).

4.7.1.3 Pervious road surfacing

Conventional bituminous materials are intended to provide an impermeable layer. This protects the underlying pavement and subgrade from rainwater. A consequence of this is that water tends to remain on the surface of the road, and arrangements must be made to minimise the amount of water at the contact point between wheel and road. If this is not done, control of vehicles will be more difficult in wet weather, and spray will reduce visibility.

An alternative approach is through the use of a wearing course in pervious macadam. This material, sometimes referred to as porous asphalt, is made with aggregate graded in such a way as to provide a high proportion of voids in the compacted material. These voids are to a large extent interconnected, and allow rain water to drain into the porous surfacing. The layers below will be of conventional materials and therefore impervious, and the normal cambered profile of the road then makes the water flow laterally through the pervious material. Positive drainage is provided at the edges of the porous layer to allow water to escape quickly from the structure of the pavement. Typical mix proportions for pervious macadam are shown in Table 4.2, and Fig. 4.7 shows a typical edge and drainage detail.

Newly laid pervious macadam will achieve the benefits of improved wheel grip and greatly reduced spray, and often a reduction in traffic noise. However, as time goes by the voids are gradually filled with dirt from the road and the benefit will reduce. Spray reduction, for example, may be as high as 95 per cent with a new pervious surface, but this figure might fall to 50 per cent or less once the material is choked. The time taken for this to happen will depend on the service conditions, and might vary between two and ten years.

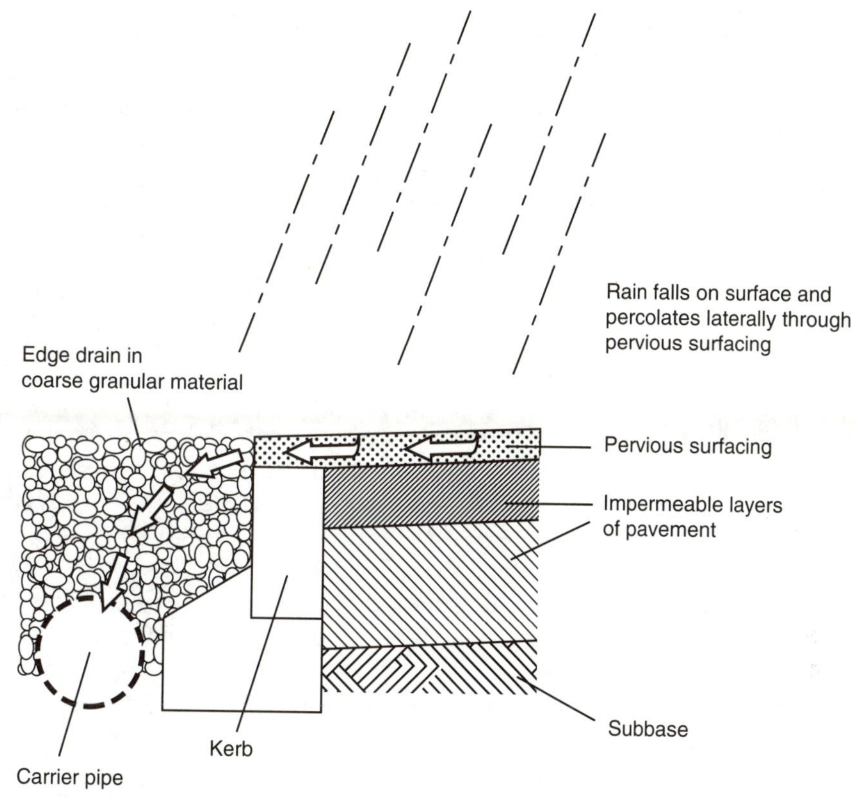

Figure 4.7 Pervious road surfacing – function and typical edge detail

Table 4.2 Pervious wearing course macadam: aggregate grading[8]

	Designation	
	20 mm size	*10 mm size*
Grading: % by mass passing		
28 mm	100	
20 mm	90–100	
14 mm	50–80	100
10 mm	–	90–100
6.3 mm	20–30	40–55
3.35 mm	5–15	22–28
600 μm	–	–
212 μm	–	–
75 μm	3–7*	3–6*

*Includes 2% by mass of total aggregate of hydrated lime.

Because of its high voids content, the material is not as durable as conventional materials, particularly asphalts, and so should not be used in areas of high traffic stress. A stiff binder is often used, to increase the strength of the material, and hydrated lime included in the aggregate to reduce the chance of binder stripping from the aggregate in the presence of water.

4.7.2 Rolled asphalt

This material consists of graded aggregate and binder, as does coated macadam, but with the differences that the grading of the aggregate may be much less continuous; that considerably more fines are present in the mixed aggregate; and that a higher proportion of the mix will be binder. The strength of the material is very sensitive to the amount of binder present and since asphalt is used in the most taxing situations it is clearly important to ensure that the correct mix proportions are achieved.

Two general types of mix design exist for rolled asphalt – the recipe mix and the design mix. These are specified in the UK in the terms of BS 594.[9]

4.7.2.1 Recipe mix

Recipe mixes quoted in BS 594 are based on tests carried out over a large number of years and represent mix designs which will be adequate in their performance in most situations. Two types of wearing course mix are presented: type F and type C.

Type F is characterised by a gap grading typical of traditional rolled asphalt wearing course mixtures, and is usually associated with the use primarily of a fine sand, although other fine aggregates complying with the grading limits are not excluded.

Type C is characterised by a coarser grading usually associated with the use of crushed rock or slag fine aggregate. However, natural sand may be used in part or in total.

The designation F reflects the finer grading of the fine aggregate in this type of mix compared with the coarser grading of the fine aggregate incorporated in the type C mix.

Binders specified in the BS are shown in Table 4.3. Recipe mixtures for roadbase, basecourse and wearing course use are shown in Tables 4.4 and 4.5.

Table 4.3 Rolled asphalt: binders recommended in BS 594

Binder No	*Type*	*Grade*
1	Bitumen	35 pen
2	Lake asphalt–bitumen	35 pen
3	Bitumen	50 pen
4	Lake asphalt–bitumen	50 pen
5	Bitumen	70 pen
6	Lake asphalt–bitumen	70 pen
7	Bitumen	100 pen

In these tables, notice that the binder content varies with the proportion of fine aggregate present in the mix, and the nature of that fine aggregate.

The intended arrangement in the proportions of a rolled asphalt is that the coarse aggregate should be set in a matrix of bituminous mortar which consists of fine aggregate (particles of sand size, between about 6.3 mm and about 75 μm), filler (particles of dust size, less than about 75 μm), and binder. The

Table 4.4 Rolled asphalt: composition of roadbase and basecourse mixes (BS 594)

	Column					
	1	*2*	*3*	*4*	*5*	*6*
Designation	50/10	50/14	50/20	60/20	60/28	60/40
Layer depth	25–50	35–65	45–80	45–80	60–120	75–150
Grading: % by mass passing						
50 mm	–	–	–	–	–	100
37.5 mm	–	–	–	–	100	90–100
28 mm	–	–	100	100	90–100	70–100
20 mm	–	100	90–100	90–100	50–80	45–75
14 mm	100	90–100	65–100	30–65	30–65	30–65
10 mm	90–100	65–100	35–75	–	–	–
6.3 mm	–	–	–	–	–	–
2.36 mm	35–55	35–55	35–55	30–44	30–44	30–44
600 μm	15–55	15–55	15–55	10–44	10–44	10–44
212 μm	5–30	5–30	5–30	3–25	3–25	3–25
75 μm	2–9	2–9	2–9	2–8	2–8	2–8
Binder content % by mass of total mixture for:						
Crushed rock or steel slag	6.5	6.5	6.5	5.7	5.7	5.7
Gravel	6.3	6.3	6.3	5.5	5.5	5.5
Blastfurnace slag: bulk density						
1440 kg/m^3	6.6	6.6	6.6	5.7	5.7	5.7
1280 kg/m^3	6.8	6.8	6.8	6.0	6.0	6.0
1200 kg/m^3	6.9	6.9	6.9	6.1	6.1	6.1
1120 kg/m^3	7.1	7.1	7.1	6.3	6.3	6.3

function of the filler is to occupy the voids between the fine aggregate particles; the function of the binder is to bind the whole together.

In deciding the mix proportions the intention is that a balance should be struck between the following needs:

- To provide a mixture which is sufficiently fluid to yield slightly under stresses arising out of changes in temperature and out of forces imposed by passing traffic, without cracking;
- To provide a mix which is sufficiently fluid to be capable of being handled and compacted satisfactorily without extraordinary effort and to accept and retain surface chippings when these are applied;
- To provide a mix which is not so fluid at ambient temperatures as to deform readily under the action of traffic, particularly on warm days, nor to allow surface chippings to be submerged in the fine aggregate/filler/binder matrix as this 'fats up' under the influence of traffic;
- To provide a mix which is not so soft and prone to abrasion that the bituminous mortar is rapidly worn away at the surface leaving the chippings to be swept away by traffic.

Table 4.5 Rolled asphalt: composition of recipe type F wearing course mixtures (BS 594)

	Column				
	1	2	3	4	5
Designation	0/3	15/10	30/10	30/14	40/14
Layer depth (mm)	25	30	35	40	50
Grading: % by mass passing					
28 mm	–	–	–	–	–
20 mm	–	–	–	100	100
14 mm	–	100	100	85–100	90–100
10 mm	–	95–100	85–100	60–90	55–85
6.3 mm	100	75–95	60–90	–	–
2.36 mm	95–100	75–87	60–72	60–72	50–62
600 μm	80–100	60–87	45–72	45–72	35–62
212 μm	25–70	20–60	15–50	15–50	10–40
75 μm	13–17	11–15	8–12	8–12	6–10
Maximum percentage of aggregate passing 2.36 mm retained on 600 μm sieve	–	18	14	14	12
Binder content % by mass of total mixture (BS 594 schedules 1A, 2A and 3A) for aggregate as shown					
Crushed rock or steel slag	10.3	8.9	7.8	7.8	7.0
Gravel	10.3	8.9	7.5	7.5	6.5
Blastfurnace slag: bulk density					
1440 kg/m³	–	9.0	7.9	7.9	7.2
1280 kg/m³	–	9.2	8.1	8.1	7.4
1120 kg/m³	–	9.4	8.3	8.3	7.6

The two significant variables which the designer of the mix may control are the hardness of the binder and the nature of the aggregate.

(a) *Hardness of the binder.* Generally, the harder the binder used the better the material will be able to resist deformation. However, harder binders demand higher rolling temperatures than do soft binders, and very hard binders have been found on occasion to be liable to brittle fracture – failure by cracking of the surface.

The grade of binder most commonly used in the UK is 50 pen, forming a material which is able to resist adequately deformation under traffic and which is not too demanding in terms of rolling temperature. In the north of the country and in lightly trafficked locations elsewhere, a 70 or 100 pen binder may be used, gaining a small reduction in the necessary rolling temperature. At sites where the finished material will be subjected to high stresses and where traffic flows are channelised – at the approaches to signal controlled junctions on heavily trafficked roads, for example – the use of a 35 pen binder will be demanded.

(b) *Nature of the aggregate.* In addition to the requirements for a successful wearing course aggregate described in Chapter 2, it is necessary to consider the effect of various types of aggregate on mix proportions.

We have seen that the recipe mix tables – Tables 4.4 and 4.5 – exhibit an apparent relationship between binder content and aggregate grading. If the binder is to perform satisfactorily, sufficient binder must be present to coat all the aggregate particles.

Coating the aggregate will demand the provision of varying quantities of binder, depending on the specific surface area of the aggregate – that is, the ratio of the exposed surface area to the volume of material considered. Determinants of specific surface will be the particle size, the shape of the particles and their surface texture. Thus large particles have less surface area per unit volume than do similarly shaped small ones; spherical particles have less surface area per unit volume than do, for example, laminars; and smooth particles have less surface area per unit volume than do more rugous ones of similar size and general shape. If an aggregate with a low specific surface is chosen, less binder will be required to coat all the particles present than would be necessary for one of high specific surface.

Thus in Table 4.5 we see that a wearing course material with a crushed rock aggregate ideally contains about 10.3 per cent binder by mass where there is no coarse aggregate in the mix (this type of material is known as a 'sand carpet'). If we increase the proportion of coarse aggregate to 40 per cent (reducing the proportion of fine aggregate and filler, and thereby the effective specific surface of the mixture) we find that the recommended binder content falls to about 7 per cent.

If, in addition to variations in aggregate size, we introduce other variations in the nature of the aggregate then we may expect further variations in the optimum mix proportions. One such variation occurs when the density of the aggregate is changed. Since we are concerned to provide a quantity of binder related to the size and shape of the aggregate particles but measure quantities by mass, a reduction in aggregate density is accompanied in Table 4.5 by an increase in the binder content expressed as a percentage of the mass of the mix; the ratio of the volumes of binder to aggregate remains the same.

A second variation may be demanded if the specific surface of the aggregate is changed. For example a fine aggregate characterised by smooth, spherical particles will have a lower specific surface than one of similar grading but characterised by more rugous, angular particles, and will therefore require less binder. The effect of this property of the aggregate on the performance of the mix is not easy to predict from a consideration of the aggregate in isolation; we therefore need to study the behaviour of the mix as a whole. A suitable means of carrying out such a study is provided by the Marshall test.

The recommendations of BS 594 for recipe mixtures are intended to provide satisfactory results in the majority of cases; but on occasion a more resilient mix may be required. In such circumstances the designer may consider the use of a harder binder, or of a mix specifically designed to suit the aggregate used at the site, rather than the assumed aggregate type implicit in the recipe method. Such a mix is known as a design mix.

4.7.2.2 Design mix

The principle upon which the design process is based is that a number of samples of binder/aggregate mixtures are prepared, using those materials to be used in the field, and varying the proportions of binder present in the mixture from sample to sample. Tests are carried out on the samples, and the mix proportion which yields the best performance in the laboratory is reported.

Details of the tests are provided in BS 598[10] to which reference should be made. The tests are complex, time consuming and expensive.

In designing an asphalt mix, we seek to provide a material which resists deformation and which is capable of being well compacted. The tests which are carried out seek to measure these characteristics of each trial mix.

In order to measure the 'compactability' of the material, a sample is prepared and compacted in a standard way into a cylindrical mould 101.6 mm in diameter and 70 mm long. The relative density of the specimen is measured by weighing in air and in water, and the compacted aggregate density is determined from a knowledge of the percentage by mass of binder present. Too low a proportion of binder will make the material hard to compact and will thus result in low measured densities (and may also cause aggregate loss or fretting at the surface in the field); too high a proportion of binder effectively displaces some aggregate from the mix, again resulting in low density. Plotting relative density against binder content will enable two 'optimum' binder contents to be estimated. Typical plots are shown in Figs 4.8 and 4.9.

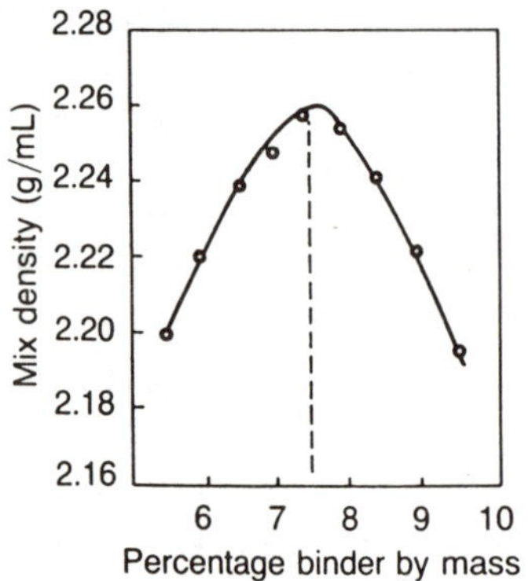

Figure 4.8 Typical density test results (a)[10]

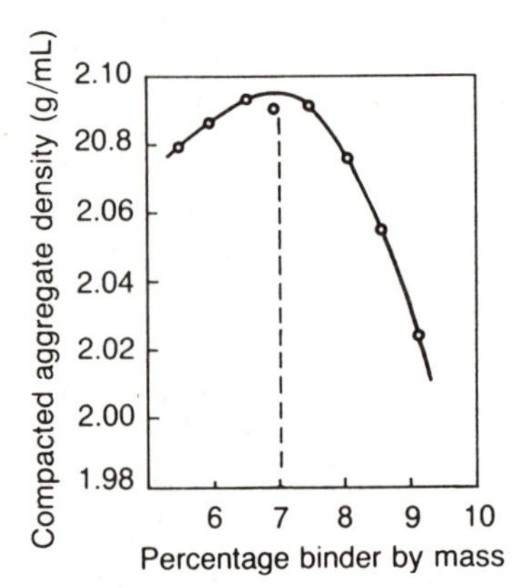

Figure 4.9 Typical density test results (b)[10]

The ability of the material to resist deformation is measured by the Marshall test, from which is obtained the stability and the flow of the mix. The test consists of placing the cylindrical sample previously prepared between two steel jaws of circular profile (matching the curved sides of the cylindrical sample) which are driven together in the test by a large force. The sample will offer some resistance to this force, which resistance may be measured by a suitably positioned transducer. The sample will first deform under loading, and then progressively fail. The jaws provide a free movement of 18.5 mm and are driven together at a standard rate of 50 mm per minute. The test happens quickly and automatic data recording is often employed.

The stability is the maximum resistance to deformation offered by the specimen, multiplied by a stability correction factor whose value depends on the original volume of the specimen.

The flow is reported as the deformation of the moulded specimen, in millimetres, at the time of maximum resistance.

Typical plots of stability and flow against binder content are shown in Figs 4.10 and 4.11.

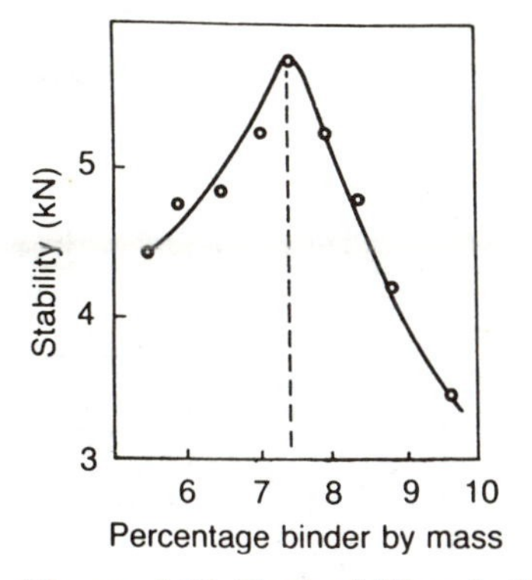

Figure 4.10 Typical Marshall test results (a)[10]

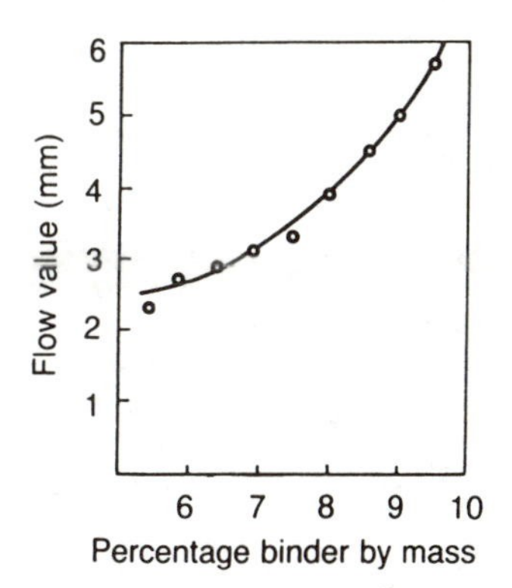

Figure 4.11 Typical Marshall test results (b)[10]

The design binder content is obtained from the average of the optimum binder contents for mix density, compacted aggregate density and stability. To this factor is added an 'addition factor' as indicated in Table 4.6. It has been found that with certain 'optimum' mixtures, fretting occurs at the surface. The addition factor is intended to avoid this and to ensure that laboratory stabilities are consistent with experience on site.

The design proposals should be verified at the construction stage. Specification requirements suggested by BS 594 are shown in Table 4.7.

Table 4.6 Rolled asphalt: design mix addition factors

Coarse aggregate content (%)	*Addition factor*
0	0
30	0.7
40	0.7
55	0

Table 4.7 Rolled asphalt: design mix specification requirements (BS 594)

Traffic (Commercial vehicles/lane/day)	*Stability (kN)*
Less than 1500	3 to 8
1500 to 6000	4 to 8
Over 6000	6 to 10

Notes:
For stabilities up to 8.0 kN the maximum flow value should be 5 mm. For stabilities in excess of 8 kN a maximum flow value of 7 mm is permissible.
These data relate to laboratory mixes, not site mixes.
Other specification requirements exist for use on trunk roads in the UK.

4.7.2.3 Reinforced asphalt

In service, asphalt and other bituminous materials may respond structurally to applied loading by fatigue cracking, or, particularly in the case of asphalts, by deformation. These failure modes arise out of the properties of the binder and result essentially from its lack of tensile strength. The performance of a bituminous pavement can be improved by the incorporation of a geogrid in the bound layer. Typical arrangements are shown in Fig. 4.12.

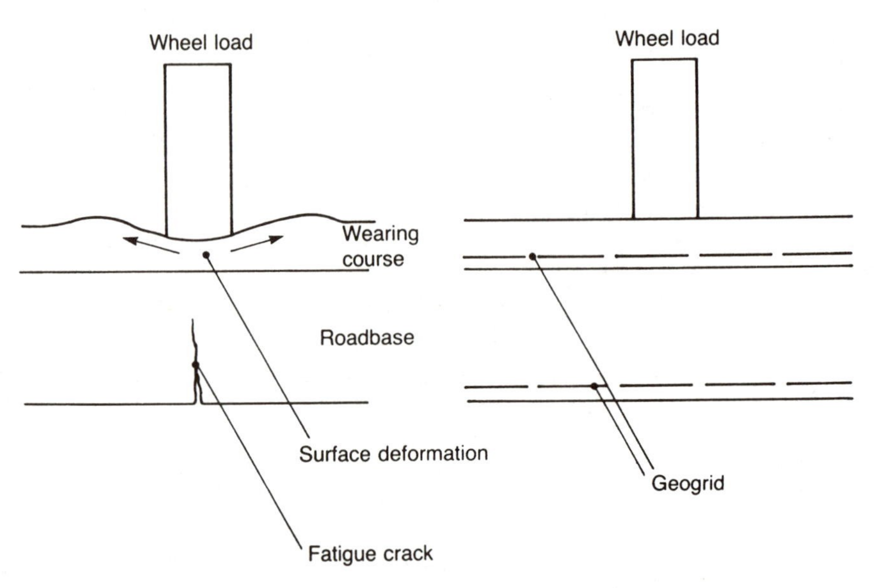

Figure 4.12 Use of a geogrid to reduce surface deformation and fatigue cracking

Placed at the bottom of the roadbase, a suitable geogrid can greatly improve the elastic properties of the bound layer and will effectively reduce the rate at which fatigue cracks are propagated, by resisting permanent strain. Once the load causing the strain is removed, the geogrid's elastic behaviour creates a tendency for cracks to close up which is not found to the same extent in the unreinforced material.

A geogrid can also have a stabilising effect when placed in the surfacing by reducing the tendency of the material to flow. Where the bituminous layer is placed as an overlay to a joined concrete pavement, the presence of a geogrid can significantly reduce reflection cracking through the bituminous layer caused by movement at the joints below. There is also evidence that the presence of a geogrid can increase the life of such pavement overlays.

In order that the unity of the bound layer is not disrupted by the geogrid, it is important that the openings in the grid are of sufficient size to allow full bonding to develop between the material on either side of the reinforcement. A typical geogrid used in such applications is formed by deforming a punched polypropylene geomembrane to form apertures 65 mm × 50 mm; the deformation has the effect of stiffening the material.

There is at present no satisfactory design method for reinforced bituminous materials.

4.7.2.4 Lean concrete

Lean concrete is a cement bound material of lower strength than pavement quality concrete. Because of its reduced strength it fails when subjected to lower stresses than those required to cause failure in PQ concrete. A consequence of this when lean concrete is used in the pavement is that the material develops frequent narrow cracks. These give an element of flexibility sufficient to warrant the material's use as part of a flexible pavement, but since the cracks are narrow they do not greatly impair the contribution the material makes to the strength of the pavement.

The behaviour of lean concrete in a pavement is not fully understood. Nevertheless a considerable body of experience of its use as a roadbase has been built up which is summarised in Fig. 4.6.

Lean concrete may either be laid wet, with a water/cement ratio comparable to that for pavement quality concrete, in which case road forms will be needed; or it may be laid dry, with a greatly reduced water content, in which case it can be handled and compacted as an unbound granular material. In either case the usual aftercare appropriate to cement-bound materials is required.

4.8 Construction of flexible surfaces

The success or failure of a flexible road depends not only on the design of the bound layers but also on the extent to which that design is realised in the construction process. Rolled asphalt particularly is not a 'forgiving' material in that errors in the construction process can be rectified only at considerable effort and expense – often only by the removal and replacement of defective work. Part 2 of BS 594 is concerned with the transport, laying and compaction of rolled asphalt.

The construction process consists of:

(1) The spreading and initial compaction of the material by a paving machine;
(2) If necessary, the spreading of coated chippings over the surface of the work;
(3) Rolling the material to compact it and to embed chippings into the surface.

It is worthwhile to consider possible pitfalls, and some of these are discussed in the following sections.

4.8.1 Delivery of material

The material may be too cold, in which case it will be impossible to compact adequately. Rolling temperatures are discussed below. Secondly, it is possible that the material will not comply with the specification. Unless the error is a gross one, visible to the eye, this is unlikely to be detected at the time of construction. The taking of adequate samples for testing is therefore of importance.

It is also possible to order the wrong amount of material to be delivered to site. Asphalt and bituminous macadam are measured by weight at the mixing plant and upon delivery to site, but by volume or its equivalent when in the works. Table 4.8 shows typical densities of compacted materials. Figures actually achieved in a particular case will depend on the aggregate used and on the degree of compaction achieved.

Table 4.8 Some typical densities of laid bituminous materials

Material	*Aggregate type*	*Typical density (kg/m³)*
Rolled asphalt wearing course	Granite	2250
Rolled asphalt base course	Gravel	2330
Bituminous macadams		
6 mm dense wearing course	Granite	2330
	Limestone	2300
	Slag	2250
14 mm close graded wearing course	Granite	2390
	Slag	2250
20 mm dense basecourse	Granite	2380
	Limestone	2350
28 mm dense roadbase	Granite	2420
	Gravel	2320
	Limestone	2360

4.8.2 Preparation of surface

If the surface on which the material is to be laid is too irregular, it will not be possible to achieve a suitably uniform surface on the finished work. Typical specification requirements are indicated in Table 4.9. Care should be taken to ensure that the surface on which the material is to be laid is within these tolerances and that it is free from loose material, standing water and anything else likely to prevent adhesion of the surfacing.

Table 4.9 Flexible pavements: surface tolerances

Top of subbase	+10–30 mm
Top of roadbase	±15 mm
Top of basecourse on major roads	±6 mm
Top of basecourse on minor roads	±10 mm
Running surface	±6 mm

4.8.3 Weather

Some types of weather are unsuitable for surfacing in bituminous materials. Examples are heavy rain, when water will be lying on the area to be surfaced, and very cold weather. BS 594 provides that 'stiff wearing course mixtures shall not be laid when the surface to be covered is below 5 °C. All laying shall cease when the air temperature reaches 0 °C on a falling thermometer but may proceed at air temperatures at or above −1 °C on a rising thermometer when the surface is dry and free from ice'.

The problem with cold weather is of course that the material cools down to a temperature below that at which it can be compacted, before enough time has elapsed for compaction to have taken place. The material is laid in a thin layer on a cold surface and is further cooled by the wind; its black colour also promotes the loss of heat. Stiff wearing course mixes should not be laid in cold weather.

4.8.4 Method of laying

By far the best method of laying bituminous materials is to use a paving machine which ensures a uniform rate of spread of properly mixed material. On very small jobs – such as patching – and in the extreme corners of intricately shaped areas, hand laying may be accepted. The greatest problem with hand laying, particularly of graded materials such as rolled asphalt and dense bitumen macadam, is that of segregation, in which the fine and coarse elements of the aggregate cease to be mixed uniformly; typically, large particles are gathered together by the laying process, forming a surface of irregular texture and less than optimum strength. This cannot readily be remedied with rolled asphalt, but with a dense coated macadam, and in circumstances where strength is not of paramount importance, the surface may be sealed by blinding with bituminous grit, closing 'hungry' areas where there was previously a shortage of fine particles, and preventing loss of the coarse material. A competent gang is capable of achieving the tolerances shown in Table 4.9 when laying by hand.

Machine laying is not entirely free of pitfalls, but the use of a well-maintained paver will overcome most. The major components of a paving machine are shown in Fig. 4.13. One error which can be avoided at the design stage is the use of oversized aggregate. This can become trapped between the screed board and the surface on which the material is being laid, and be dragged forward by the machine to cause characteristic scarring in the new work.

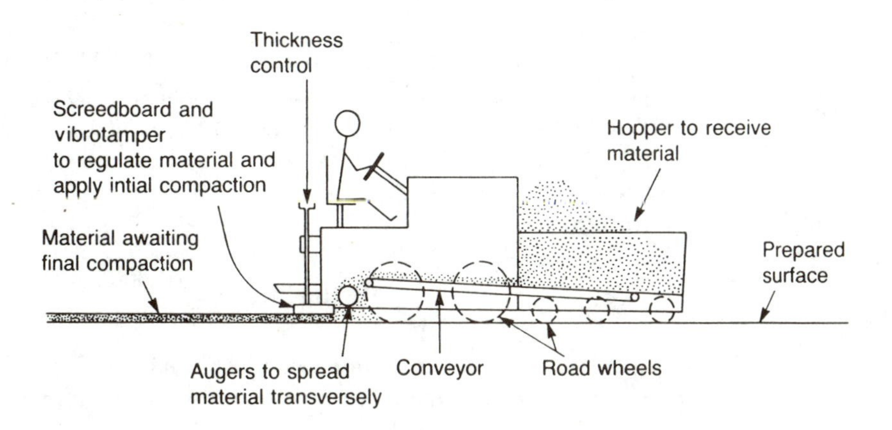

Figure 4.13 Major features of a paving machine

4.8.5 Falls and drainage

Since the material cannot be laid to a perfectly planar profile, it is important to provide adequate gradients in the finished surface to allow surface water to drain off the slightly undulating surface of the pavement. If this is not done, pools of standing water will form which are a potential hazard to traffic and which may penetrate the pavement. This is discussed at some length in Chapter 7. Typical specification requirements are a minimum transverse gradient of 2 per cent, and a minimum longitudinal gradient against a kerb of 0.8 per cent. To promote free drainage, it is common practice to keep the channels free of chippings for a width of about 150 mm.

4.8.6 Surface texture

Chippings are commonly applied to provide enhanced surface texture, and are placed on the surface of the material after laying but before compaction. It is necessary to ensure that the chippings are uniformly spread, spread at the correct rate, and rolled in far enough to enable the surfacing to hold them in place, but not so far as to cause them to be submerged and have no effect.

Uniformity of spreading is best achieved by the use of a mechanical chipping spreader. Otherwise it is possible to achieve satisfactory results by hand – but much more difficult! Visual checks can be made at the time of laying.

Rate of spread was once determined from a consideration of many variables. This was seldom satisfactory. Current practice is to determine the rate of spread required to achieve full coverage, with the chippings laid shoulder to shoulder, and to set as a target rate of coverage a proportion, usually 70 per cent, of this rate; with the proviso that a relaxation may be possible if adequate skidding resistance – and therefore texture depth – can be achieved.

4.8.6.1 Texture depth

If chippings are lost in the surfacing, are not big enough, are not spread at a sufficient rate, or are lost from the surface under the action of traffic, inadequate texture depth will be achieved. This will result in poor skidding resistance. A typical target texture depth is 1.5 mm as measured by the sand patch test or other means.

The sand patch test is fully described in BS 598. It consists of taking a standard volume (50 ml) of sand with a specified grading, spreading it on the surface in a circular patch so that the surface depressions are filled with sand to the level of the peaks, measuring the diameter in millimetres of the circle thus formed – denoted by D – and calculating the texture depth from the relationship

$$\text{Texture depth} = \frac{63\,660}{D^2}$$

An alternative to the sand patch test which can assess surfaces more thoroughly and quickly is offered by the mini texture meter, developed by the TRRL. Pulses of laser-generated light are projected onto the pavement surface and an array of photo-sensitive diodes detect the reflected light, enabling the scatter of the light and hence the surface texture to be assessed. The mini texture meter is intended to provide continuous measurement while being pushed over the surface by hand and includes a microcomputer to interpret and accumulate the data.

A larger machine known as the high-speed texture meter is intended to be vehicle mounted and generates light pulses at such a rate that surface texture measurements can be made at speeds of up to 100 km/hr.

4.8.7 Rolling

Sufficient rollers should be present to keep up with the paving machine. Each should be a dead-weight machine weighing not less than 8 t, or an equivalent vibratory machine. Minimum rolling temperatures are set out in Table 4.10. It is important that the material is fully compacted before the temperature of the material falls below these limits. Purpose made thermometers are available to measure the temperature of thin layers of material *in situ*.

Table 4.10 Rolled asphalt: minimum rolling temperatures[9]

Pen grade of binder	*Minimum rolling temperature (binder viscosity of about 300 poise)*
35	90 °C
50	85 °C
70	80 °C
100	75 °C

4.8.8 Surface regularity

In addition to the requirements shown in Table 4.9 it is important to ensure that the surface of the pavement is properly regular. Longitudinal regularity is measured by the rolling straight edge, a many-wheeled trolley arranged so that vertical movement of individual wheels relative to the whole assembly can be identified. A simple straight edge, 3 m long, is used for transverse measurements of regularity, and, in situations where the use of the rolling straight edge is impractical, for longitudinal measurement.

Typical[11] specification requirements in respect of irregularities measured by the rolling straight edge are shown in Table 4.11. Deviations measured beneath the 3 m straight edge should not exceed 3 mm.

Table 4.11 Maximum permitted number of surface irregularities – public roads

	Roads and associated surfaces				*Lay-bys, service areas and surfaces immediately beneath the wearing course*			
Irregularity	4 mm		7 mm		4 mm		7 mm	
Length (m)	300	75	300	75	300	75	300	75
Principal roads	20	9	2	1	40	18	4	2
Minor roads	40	18	4	2	60	27	6	3

4.8.9 Laying asphalt reinforcement

The provision of a geogrid reinforcement in a bituminous pavement is made complicated by the need for the paver to traffic the geogrid, and by the disturbing effect of the augers used in the machine to spread material laterally. Additionally one must ensure that the geogrid is correctly oriented and positioned within the bound layer. Debonding may occur at wrinkles when the geogrid shrinks under the action of the hot asphalt.

After cleaning and preparing the substrate to receive the bituminous material, the geogrid is unrolled over the surface, tensioned, and firmly fixed in place. Where the substrate is bound, the fixing may be by clips nailed to the surface; elsewhere other arrangements are necessary. It is possible to lay the grid to radii down to about 300 m; tighter curves must be negotiated in a series of short straights. The geogrid is then protected from disturbance by the paver by placing on it a thin layer of bituminous material – typically a surface dressing. Where this is too expensive, as can be the case in some maintenance projects, a coated macadam may be hand laid in a thin layer over the reinforcement. Paving then continues in the normal way.

Additional plant required for the process includes one or more tensioning beams (fitted with hooks to engage in the geogrid), and a nail gun or other anchoring device. The additional cost of the work is generally more than offset by the savings to be made from increased pavement performance.

Revision questions

1 (**a**) There are three modes of structural failure associated with bituminous pavements. With the aid of diagrams, describe each mode of failure and explain the techniques which can be used to delay such failures.

(**b**) Explain the reasons for using composite bituminous roadbase construction for pavements designed to carry very large volumes of traffic.

2 Compare the semi-empirical design method of LR 1132 with an analytical method of pavement design. Explain how the two methods differ and why the former is not always adequate for the preparation of pavement designs.

3 Describe two alternative methods by which rolled asphalt may be specified, with particular reference to BS 594. Which is recommended for use on heavily trafficked sites? Why?

4 If you were to supervise a construction gang laying a rolled asphalt wearing course, explain the various tests and procedures you would adopt to ensure the completed work was of an acceptable standard. Indicate why each of the tests and procedures is relevant to the process of quality control.

5 In the analytical design of a bituminous pavement, design factors include various properties of the bituminous materials, and considerations which depend on the site and the desired life of the pavement. What are these considerations? Why are they important?

References and further reading

1 'Road note 29 – A guide to the structural design of pavements for new roads', Third edition, Department of the Environment, HMSO, 1973.
2 Brown, S. F., 'An introduction to the analytical design of bituminous pavements', University of Nottingham, 3rd edition, 1983.
3 Cooper, P. E. and Pell, P.S., 'The effect of mix variables on the fatigue strength of bituminous materials', TRL LR633, 1974.
4 Bonnaure, F., Gest, G., Gravois, A. and Uge, P., 'A new method of predicting the stiffness of asphalt paving mixtures', *Proc. Assn. of Asphalt Paving Tech.*, Vol. 46, 1977, pp. 64–104.
5 Ullidtz, P., 'A fundamental method for the prediction of roughness, rutting and cracking in asphalt pavements', *Proc. Assn. of Asphalt Paving Tech.*, Vol. 48, 1979, pp. 557–86.
6 Robinson, R. G., 'A model for calculating pavement temperatures from meteorological data', TRL LR44, 1985.
7 Powell, W. D., Potter, J. F., Mayhew, H. C. and Nunn, M. E., 'The structural design of bituminous roads', Transport Research Laboratory, Laboratory Report 1132, 1984.
8 'Coated macadam for roads and other paved areas', BS 4987, BSI, 1988.
9 'Hot Rolled Asphalt for roads and other paved areas', BS 594, BSI, 1992.
10 'Sampling and examination of bituminous mixtures for roads and other paved areas', BS 598: Parts 104–109, BSI, 1989/1990.
11 'Specification for Highway Works', Department of Transport, HMSO, 1991.
12 Departmental Standard HD 14/87 'Structural Design of New Road Pavements', Department of Transport, 1987.

5. Concrete and composite pavements

5.1 Introduction

A pavement's job is to protect the subgrade from the damaging effects of traffic, and to provide a robust surface with suitable characteristics for the passage of traffic. Rigid pavements are those which discharge these functions by means of large concrete slabs, at least one traffic lane wide and of proportionate length, and have the general characteristic that they are able to span over minor irregularities in the subgrade.

In the decades following the Second World War concrete pavements were widely used throughout the UK. Expected benefits were durability and long life. These remain achieveable, but maintanence problems associated with concrete pavements have often been significant with the result that the material is now seldom used alone for new road construction. Sudden major failures have occurred on motorways and trunk roads in the UK and elsewhere. Composite pavements, using concrete and bituminous materials, remain a frequent choice.

However, in some specialist circumstances the wholly concrete pavement continues to provide the best option. The material's ability to withstand the action of hydrocarbons such as diesel fuel, and its long life when undisturbed, ensure that it will retain a niche application.

5.2 Design of jointed concrete pavements

Generally a rigid pavement will consist of a foundation, as discussed in Chapter 3, and a slab in pavement quality concrete. Various attributes of the material dictate the general form of the pavement: the strength of the material, in tension and in compression; its fatigue characteristics; and the changes which occur in concrete as it sets.

Concrete is a brittle material which responds to excessive stresses by developing cracks. The aim of the designer is to ensure that a sufficient depth of concrete is provided to prevent premature failure as a result of the action of traffic, and that suitable measures are taken to prevent the slab being damaged by other causes. The relationship between the two design considerations is complex and is therefore considered here first in outline, and then by a detailed examination of each variable in the design process.

5.2.1 General principles of design

The designer can control the mix proportions and constituent material of the concrete, the dimensions of each slab formed of the concrete, the way in which the joints between the slabs are formed, and the support conditions beneath the slab.

The mix proportions and the choice of constituent materials will influence the strength of the concrete, and its resistance to wear and to damage by climatic and chemical attack. Mix proportions also affect the way the concrete behaves early in its life – particularly the amount of shrinkage while the material is curing. General properties of concrete are discussed in Chapter 2.

Slab dimensions are very important. The thickness of the slab is a major determinant of its ability to support repeated traffic loads. As the temperature of a mature slab changes, or as a new slab cures and shrinkage occurs, tensile or compressive forces will be set up in the slab and the magnitude of these will be a function of its plan dimensions. Joints may be provided to relieve these forces, and reinforcement may be used to increase the concrete's ability to withstand them. Joints must be formed in such a way as to remain effective for a reasonable period without the need for expensive maintenance.

Support conditions beneath the slab also influence its life – well supported slabs last longer than otherwise similar poorly supported slabs.

5.2.2 Slab design – thickness

In preparing any design, it is advantageous to be aware of the state which one's design seeks to prevent – in other words, a definition of failure is useful. Such a definition enables performance to be monitored and the effectiveness of different designs to be compared. In the case of a rigid pavement, a suitable performance yardstick is offered by the development of cracks visible at the surface. Such cracks may be measured in terms of their width, their length or their frequency of occurrence. Table 9.7 (p. 226) is relevant here. Wide cracks, in excess of 1.5 mm, demand slab replacement as the only effective treatment and it is the development of wide cracks across the full width of a slab which is therefore accepted as a reliable measure of pavement failure.

The design of a concrete road slab is a complex problem by virtue of the nature of the support given to the slab by the pavement foundation, and by virtue of the infinite variety of load combinations that can be applied to the surface. Mathematical models of pavement behaviour have been developed[1] but these have been complex and have not fully reflected service conditions, particularly in respect of loading combinations. Designs have therefore generally been based on empiricism.

Mayhew and Harding[2] took research data from experimental pavements at various sites in the UK with service lives of up to 30 years and over 30 msa. Two equations were derived relating pavement life to a number of variables identified as significant. The best models of the available data are:

for unreinforced concrete pavements

$$\text{Ln}(L)=5.094\ \text{Ln}(H)+3.466\ \text{Ln}(S)+0.4836\ \text{Ln}(M)+0.08718\ \text{Ln}(F)-40.78 \qquad [5.1]$$

in which Ln is the natural logarithm

L is the cumulative traffic, in msa, that can be carried before failure
H is the thickness, in mm, of the concrete slab
S is the 28 day mean compressive strength, in MPa, of cubes of the pavement concrete
M is the equivalent modulus, in MPa, of a uniform foundation giving the same slab support as the actual foundation
F is the percentage of failed slabs, failure being defined as one of:

- a medium crack crossing the slab longitudinally or transversely
- longitudinal and transverse cracks of medium width both starting from an edge and intersecting
- wide corner cracking
- pumping at a joint or edge
- replaced or structurally repaired bays;

and for reinforced concrete pavements

$$\text{Ln}(L)=4.786\ \text{Ln}(H)+1.418\ \text{Ln}(R)+3.171\ \text{Ln}(S)+0.3255\ \text{Ln}(M)-45.15 \qquad [5.2]$$

in which L, H, S and M are as before and
R is the amount of high yield steel reinforcement, in mm^2/m, present in the reinforced concrete slab, measured in terms of the cross-sectional area of longitudinal reinforcement per unit width of slab. Long mesh reinforcement to BS 4483 is generally used in the UK. Standard mesh sizes have $R=385$, 503 or 636 mm^2/m.

In both of these relationships, an estimation of the equivalent foundation modulus is required. Table 5.1 provides typical values for this quantity, derived by Mayhew and Harding for various foundations for concrete pavements.

The slab thickness in equations [5.1] and [5.2] represents that which with a slab whose surface is wholly trafficked has a 50 per cent probability of achieving the indicated service life. The designer may wish to vary these two parameters, in that a higher probability of achieving the stated life may be required – typically an 85 per cent probability is considered acceptable – and the surface of the slab may not be wholly trafficked.

Designers often include a margin, perhaps a metre wide at each side of the road, whose purpose is primarily related to traffic requirements – enabling vehicles stopped on the road to pull over from the main traffic stream to some extent, providing a route for cyclists and so on. Such a marginal strip can, if constructed as part of the whole pavement and tied to it, increase the effective strength of a concrete pavement. Alternatively, for a given design life, the thickness of the pavement may be reduced where such a tied shoulder is

Table 5.1 Typical equivalent foundation moduli

Subbase						*Subgrade*		
Upper Layer			*Lower Layer*					
Type	*Depth* (mm)	*Mod.* (MPa)	*Type*	*Depth* (mm)	*Mod.* (MPa)	*CBR* (%)	*Mod.* (MPa)	*Eq. Found Mod.* (MPa)
Granular			Capping	600	70	1.5	23	68
Type 1	150	150		350	70	2	27	65
	225	150	None			5	50	89
Lean				600	70	1.5	23	261
concrete			Capping	350	70	2	27	268
(C10)	150	28,000		150	70	5	50	358
			None			15	100	683
Lean				600	70	1.5	23	277
concrete			Capping	350	70	2	27	285
(C15)	150	35,000		150	70	5	50	383
			None			15	100	732

provided – although, of course, often at the cost of extra land take as well as adding to the construction cost.

Figure 5.1 indicates the variations in slab thickness recommended to make allowance for tied shoulders at least one metre wide. Where a tied shoulder is provided the slab thickness correction indicated by curve (a) should be subtracted from the slab thickness indicated by equation [5.1] or [5.2].

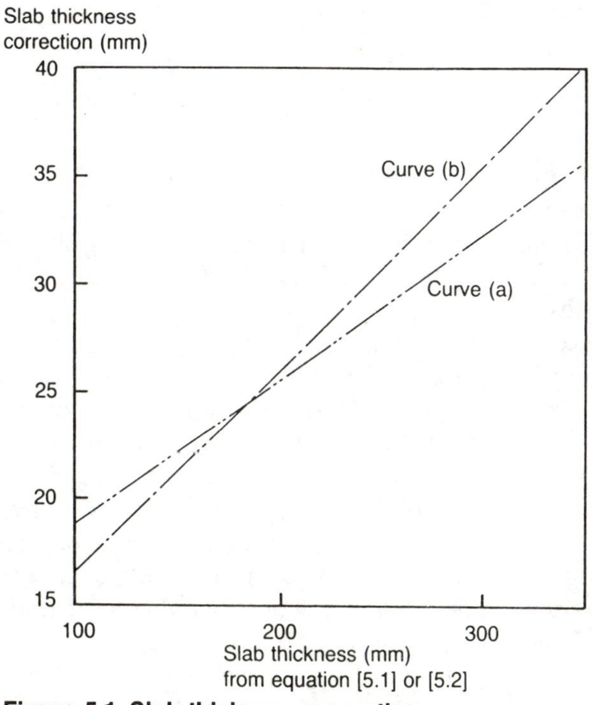

Figure 5.1 Slab thickness corrections

Often the designer seeks a probability of survival of greater than 50 per cent – typically, an 85 per cent probability is required. Figure 5.1 also indicates, by curve (b), the additional thickness which should be provided in addition to that indicated by equations [5.1] or [5.2] to achieve this increased probability of survival. These two equations may readily be rearranged as required and solved, and any necessary corrections applied. Figure 5.2 indicates designs deduced from these equations with the following 'standard' parameters:

S, 28 day mean compressive strength equal to 48.2 MPa, the target strength associated with a C40 mix and a standard deviation of 5 MPa (see Chapter 2)

M, equivalent foundation modulus, as shown in Table 5.1 for some of the commonest foundation types

R, the amount of reinforcement, corresponding to the standard meshes mentioned above

F, the percentage of unreinforced slabs failing, equal to 30

85 per cent probability of survival.

These charts indicate the sensitivity of the design method to changes in some of the variables and, in particular, indicate the increase in pavement life that may be expected as a result of the use of reinforcement. Pavements which do not include lean concrete or other bound material in the foundation demonstrate a threefold increase in the predicted pavement life when a standard C636 mesh is included in the design, in comparison with that corresponding to the unreinforced slab. Cost differentials vary from one case to another, and the designer should consider the implications of each combination of slab depth and quantity of reinforcement which will satisfy the design requirements.

5.2.3 Slab design – joints and joint layout

As the concrete in a newly made slab sets, the chemical changes that take place result in the evolution of modest amounts of heat, which gives rise to no problem in pavement construction since this heat is readily dissipated. There is also a small but significant amount of shrinkage in the concrete, whose effect is to set up tensile stresses which the immature concrete may be unable to withstand.

Once the slab is in service it will be subject to variations in temperature. These may be due to the slab cooling and warming gradually – owing, for example, to changes in the season – and will result in a general contraction or expansion in the slab. At other times the variations may be more local – due to warm sunshine on the upper face of a cold slab – resulting in differential thermal expansion of the slab and consequent warping.

The effect of these changes is that a continuous slab of concrete of the thickness and with the reinforcement indicated in Fig. 5.2 is likely to fail prematurely, not due to the action of traffic but because of internal stresses. The designs suggested by Fig. 5.2 must therefore be modified in some way.

There are two general approaches to the problem. One response to the situation in which the slab is likely to be disrupted by internal forces is to

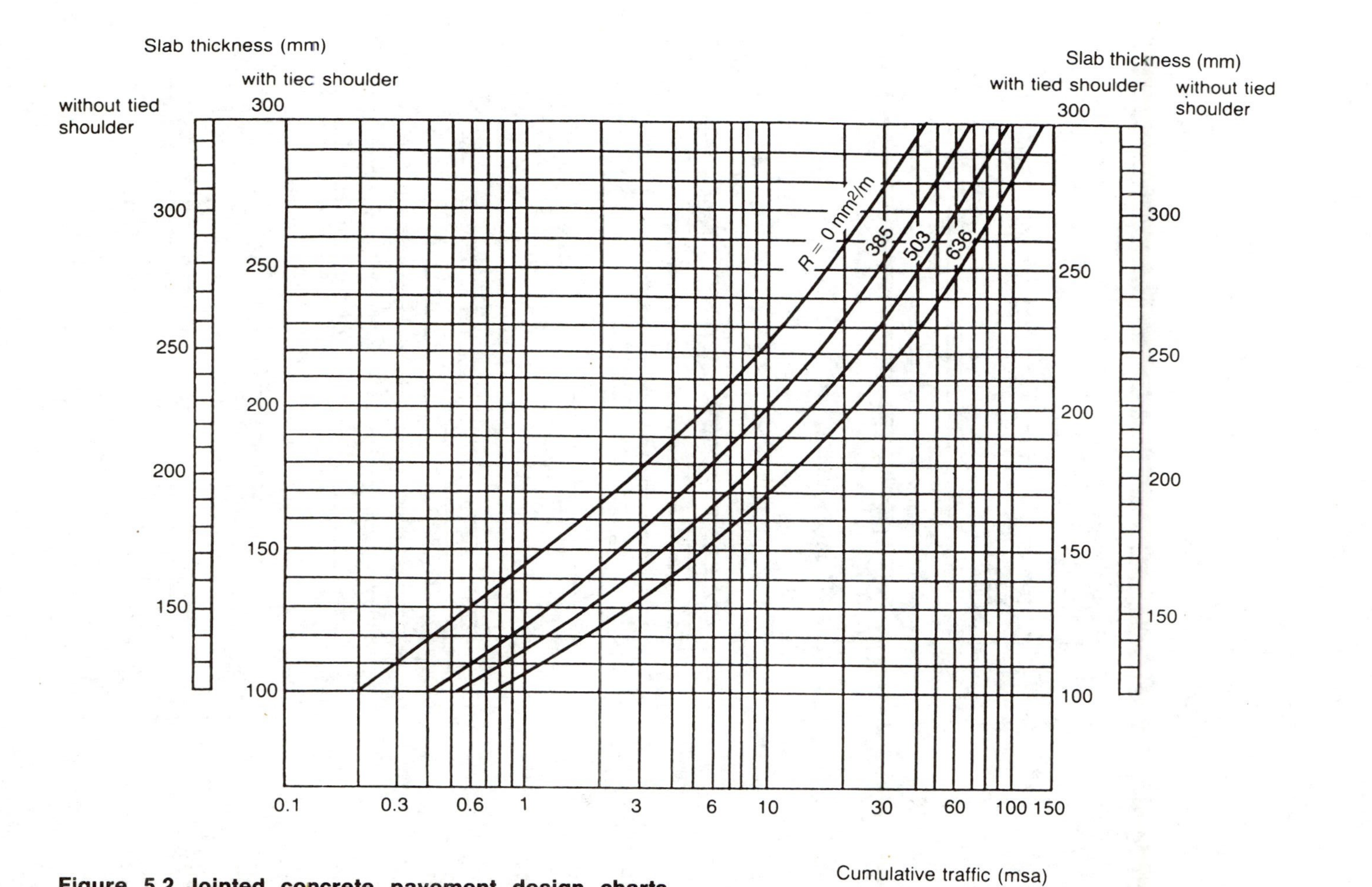

Figure 5.2 Jointed concrete pavement design charts
(a) M=68 MPa

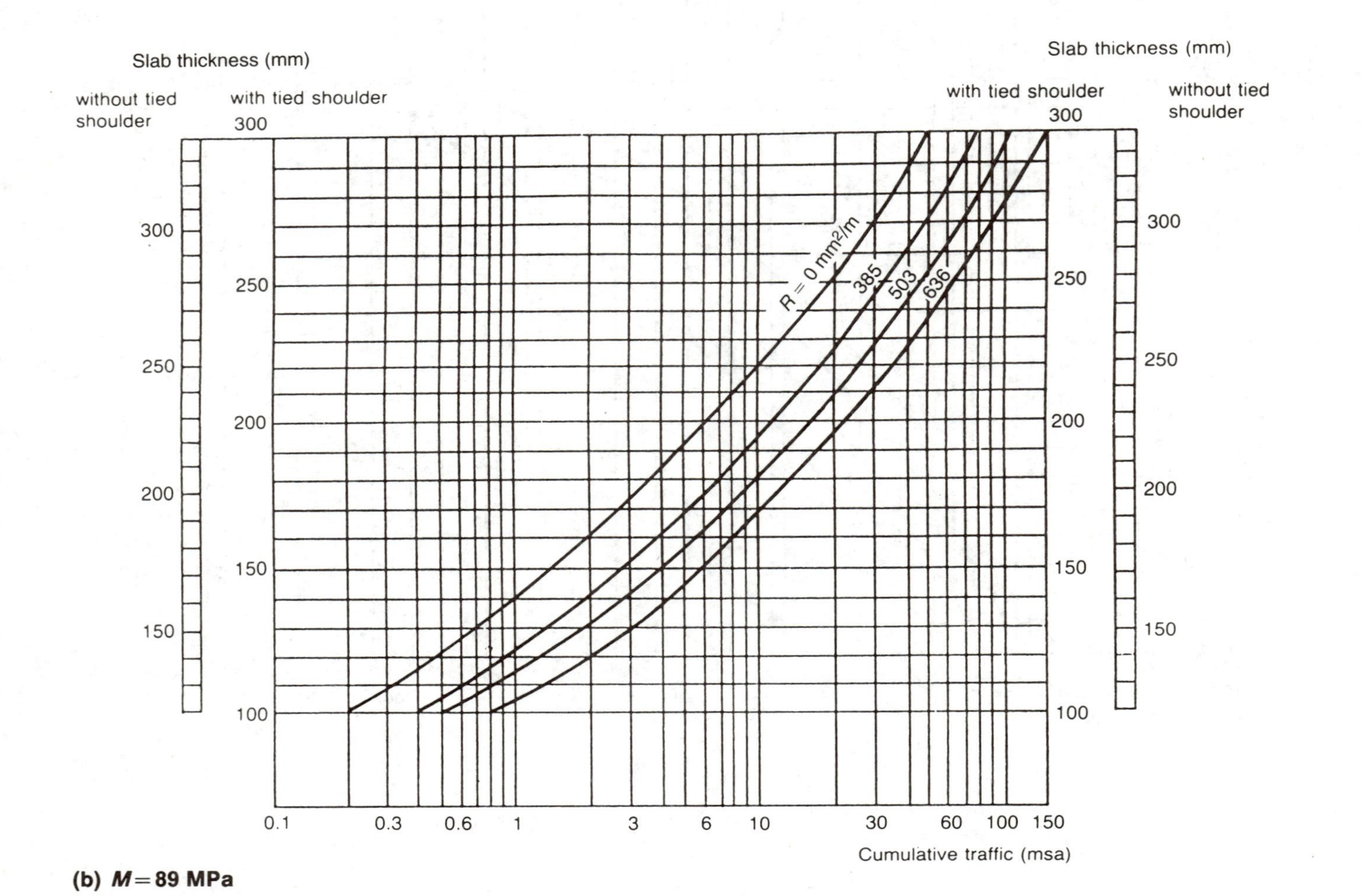
Slab thickness (mm)
without tied shoulder
with tied shoulder
300
250
200
150
100
Slab thickness (mm)
with tied shoulder
without tied shoulder
R = 0 mm²/m
385
503
636
0.1
0.3
0.6
1
3
6
10
30
60
100
150
Cumulative traffic (msa)

(b) $M = 89$ MPa

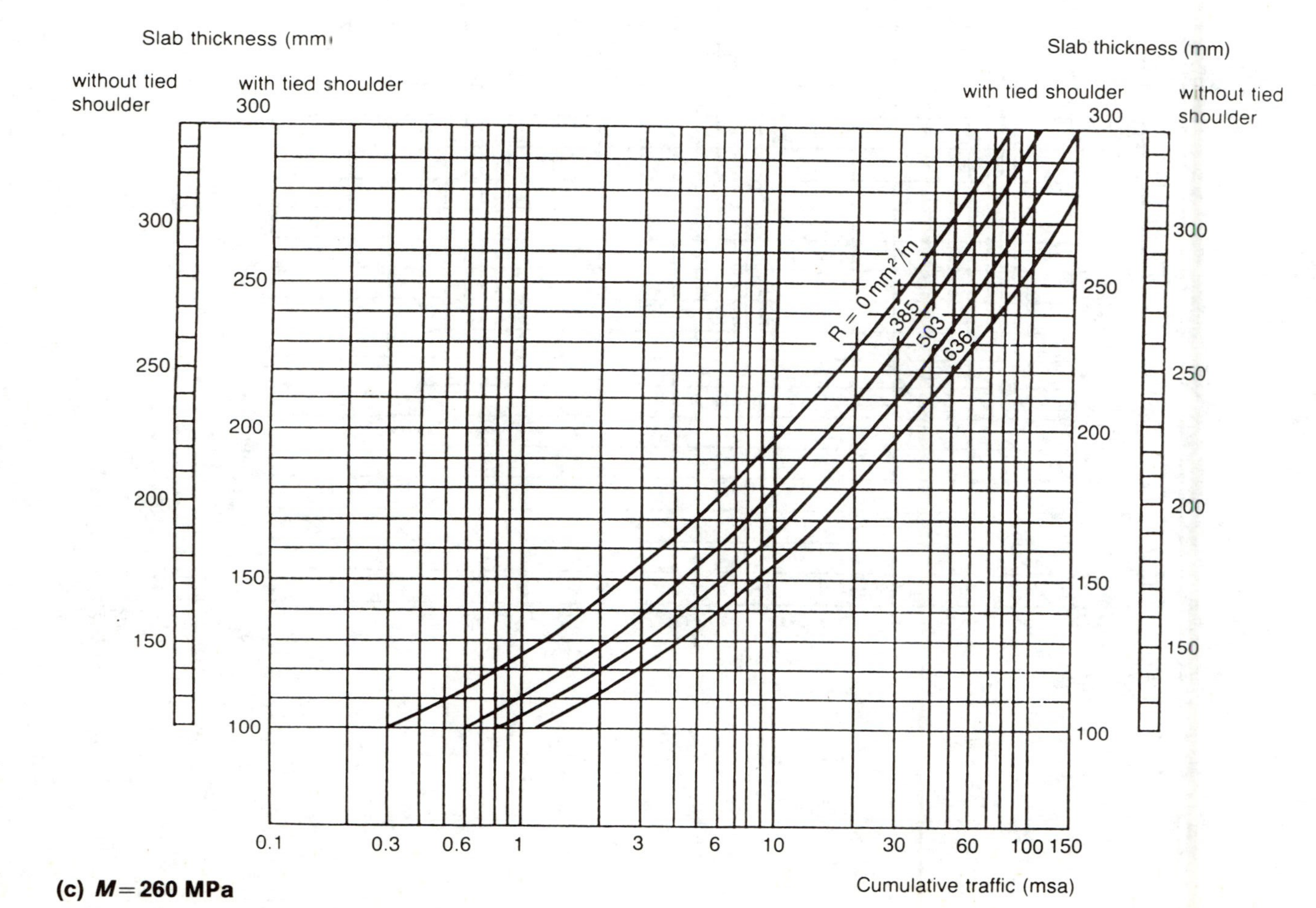
Slab thickness (mm)
without tied shoulder
with tied shoulder
Slab thickness (mm)
with tied shoulder
without tied shoulder
R = 0 mm²/m
385
503
636
300
250
200
150
100
0.1
0.3
0.6
1
3
6
10
30
60
100 150
Cumulative traffic (msa)

(c) $M = 260$ MPa

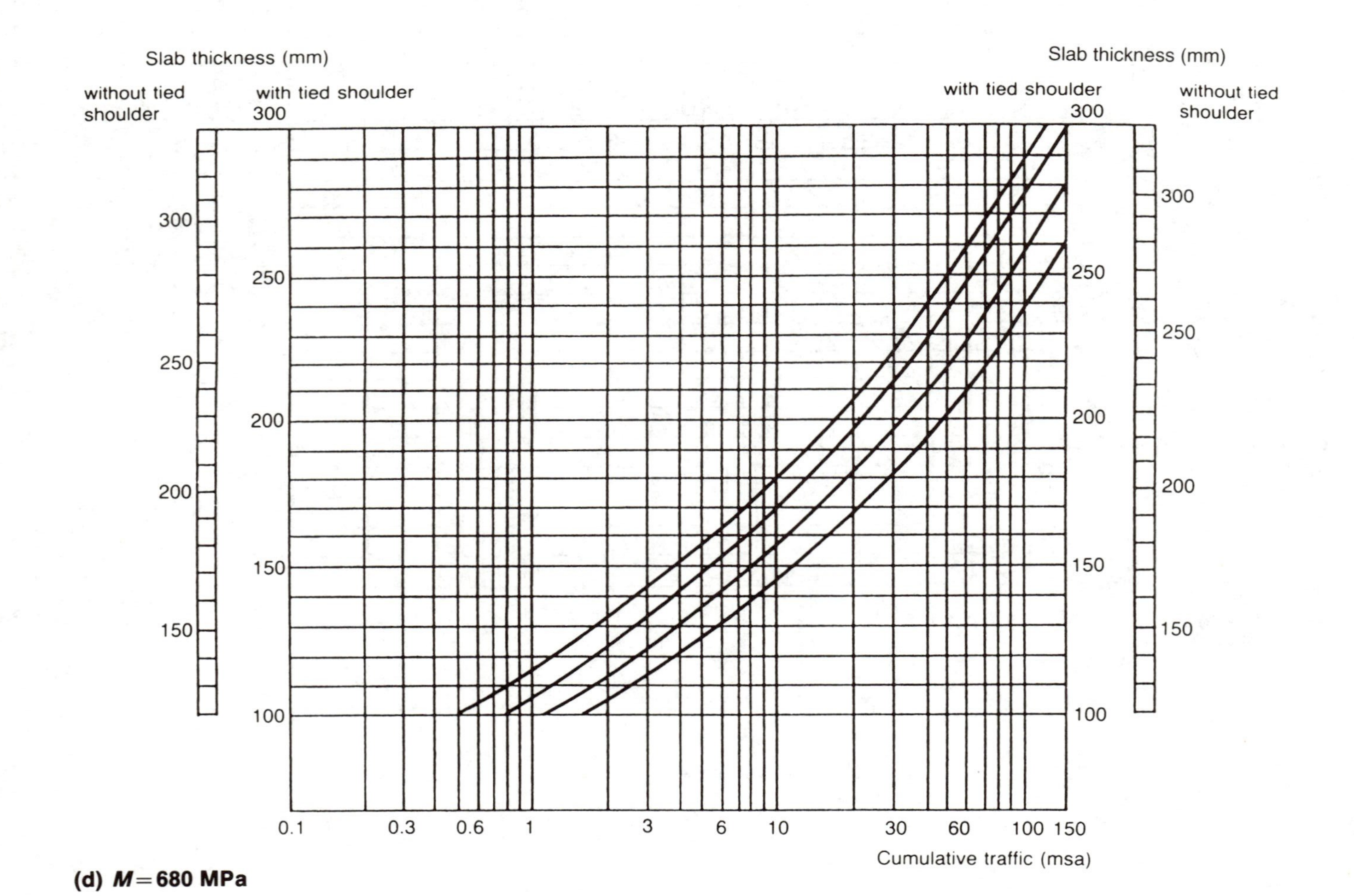
Slab thickness (mm)
without tied shoulder
with tied shoulder
300
250
200
150
100
Slab thickness (mm)
with tied shoulder
without tied shoulder
0.1
0.3
0.6
1
3
6
10
30
60
100 150
Cumulative traffic (msa)

(d) *M*=680 MPa

strengthen the slab, generally by the addition of further reinforcement, so that it is better able to withstand these forces. In such a pavement the slab and reinforcement are continuous, giving rise to the generic name of Continuously Reinforced Concrete Pavement (CRCP). This arrangement, also used in composite pavements, is discussed later.

The other possible response is to reduce the forces acting at any particular point, by dividing the pavement into a series of slabs and providing movement joints between these to allow stress release. Such an arrangement is known as a Jointed Concrete Pavement (JCP).

In the design of a JCP two further issues must be resolved – the nature of the joints, and their disposition throughout the work.

5.2.3.1 *Joint types*

For the sake of convenience, joints are often broadly categorised by their orientation in the finished work. Thus transverse joints are aligned across the pavement, at right angles to its centre line, and are provided to release stresses whose greatest effect is longitudinal – such as shrinkage and thermal contraction and expansion. Longitudinal joints should allow for effects most marked across the width of the pavement, such as warping, and angular movement due to variations in the support conditions which may arise as the moisture content of the subgrade varies at the edges of the pavement but remains virtually constant below the centre line. It is often also necessary to provide joints in both directions to facilitate construction.

5.2.3.2 *Transverse joints*

There are four types of transverse joints in general use – contraction, expansion, warping and construction joints.

Contraction joints provide stress release by allowing the adjacent slabs to contract and thereby reduce tensile stresses in the concrete. The joint must therefore be capable of opening, to allow this movement, and of subsequently reclosing to allow thermal expansion to the original slab dimensions. It must allow neither differential vertical movement between adjacent slabs, nor water penetration into the pavement foundation. Contraction joints are formed by reducing the thickness of the slab where required so that the subsequent stress concentration will cause a crack to form in the appropriate place. This thickness reduction may be achieved either by means of a crack inducer together with a groove formed in the surface, as in Fig. 5.3, or by omitting both of these and instead cutting a groove in the surface with a power saw after the initial set has developed. In either case, the combined reduction in the slab depth should be of the order of 25–35 per cent at the joint. The dowel bar provides the necessary vertical shear strength across the joint and is debonded on one side of the joint to allow longitudinal movement. The seal at the top of the joint excludes dirt and water.

Expansion joints, as might be expected, allow the adjacent slabs to expand beyond their original sizes. They differ from contraction joints in that the discontinuity in the slab is performed and in that a strip of compressible

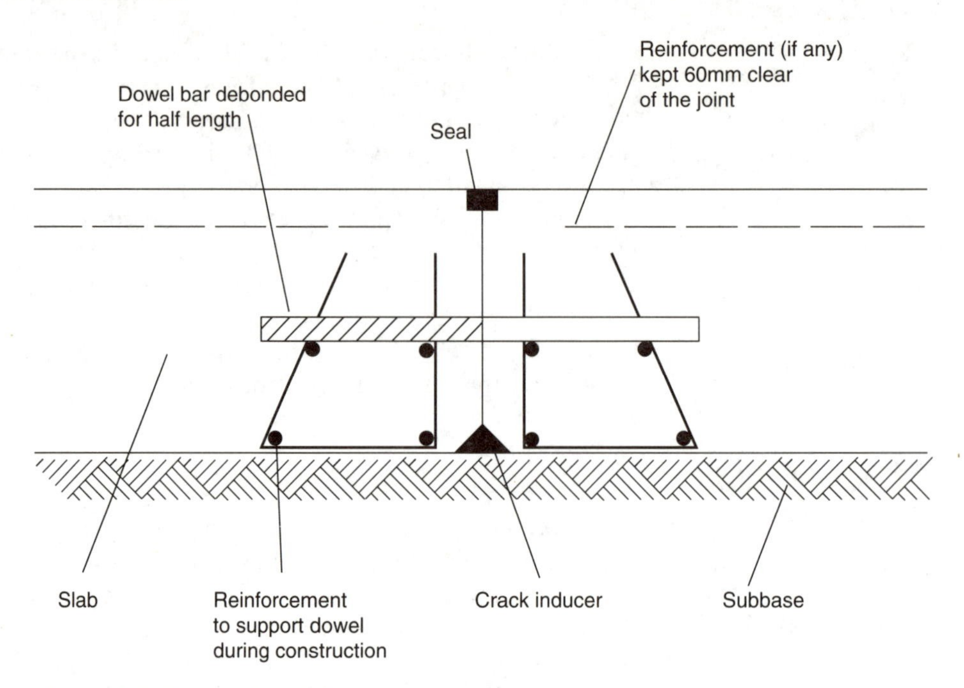

Figure 5.3 Contraction joint

material (various proprietary products are available) is included in the joint to allow the subsequent expansion of the adjacent concrete. These differences demand the interruption of concreting work or the making of special arrangements and this, together with the extra material cost, means that expansion joints are costly. Expansion joints can also function as contraction and warping joints. A typical detail is shown in Fig. 5.4(a). When concreting takes place in the summer, it is often acceptable to provide contraction joints in lieu of expansion joints.

Poorly maintained expansion and contraction joints are a major cause of failure in JCPs. If the joint opens more than the flexible seal can accommodate, detritus will enter and prevent the joint re-closing. Figure 5.4(b) shows this, and Fig. 5.4(c) shows a blow-up failure caused by thermal expansion of a JCP whose expansion and contraction joints had not been kept clear of detritus.

Warping joints are intended to allow slight angular movement, relieving stresses due to differential expansion and to variations in support conditions. They are no more than a sealed discontinuity in the concrete, the sides of which are held together by tie bars, as shown in Fig. 5.5. Transverse warping joints are necessary only in plain concrete pavements.

Normally, construction should be arranged so that work ends each day at a contraction or expansion joint. This is not always possible and, if an additional expansion or contraction joint cannot be provided, a construction joint may be used. The intention here is to allow no relative movement across the joint; a typical detail is shown in Fig. 5.6. This joint type is most often used when working next to previously constructed work where an additional transverse joint cannot be provided. Transverse joints intended to allow movement should always be continuous across the full width of the pavement.

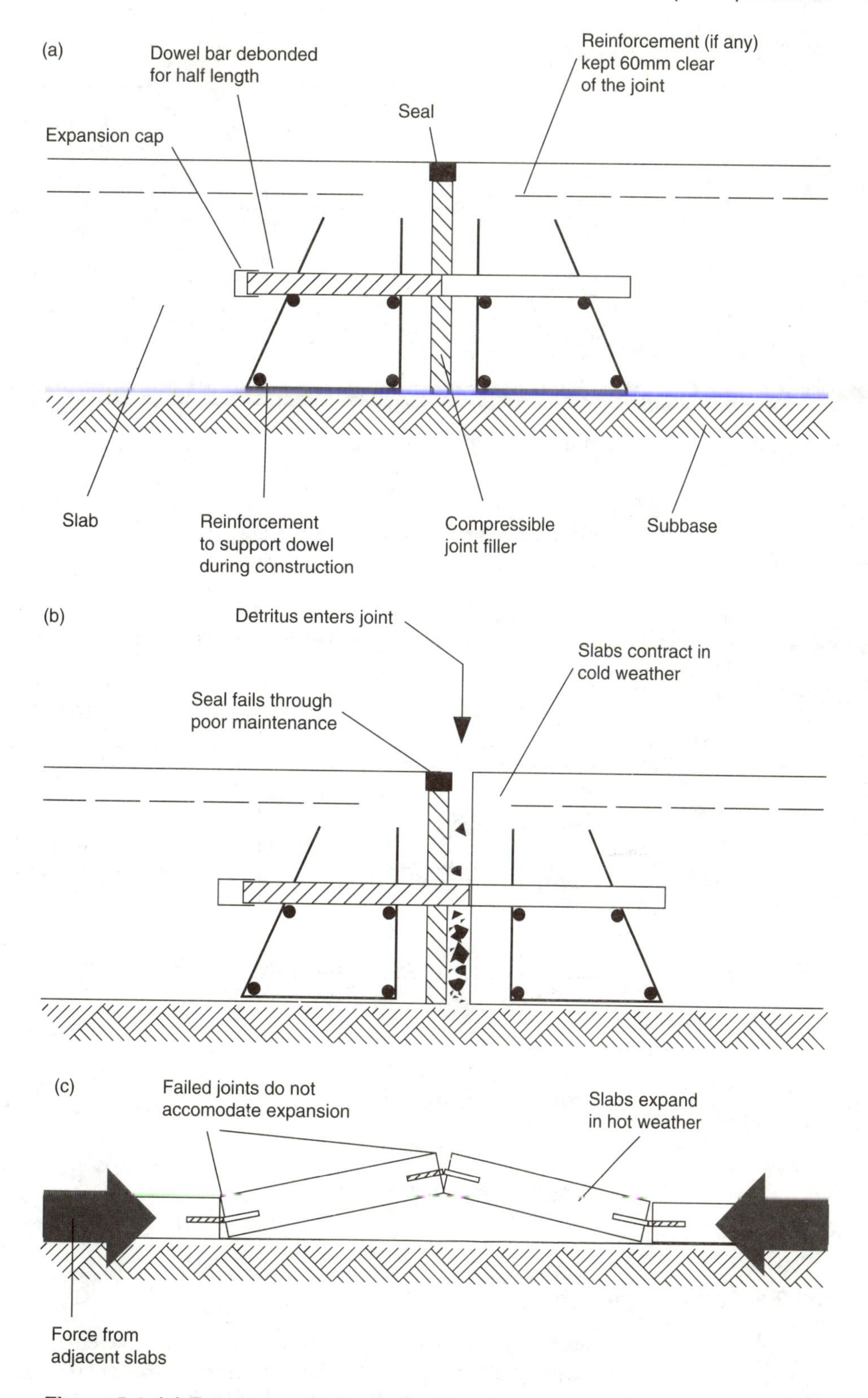

Figure 5.4 (a) Expansion joint details. (b) Failure mechanism at joint. (c) Blow up or failure of jointed concrete pavement

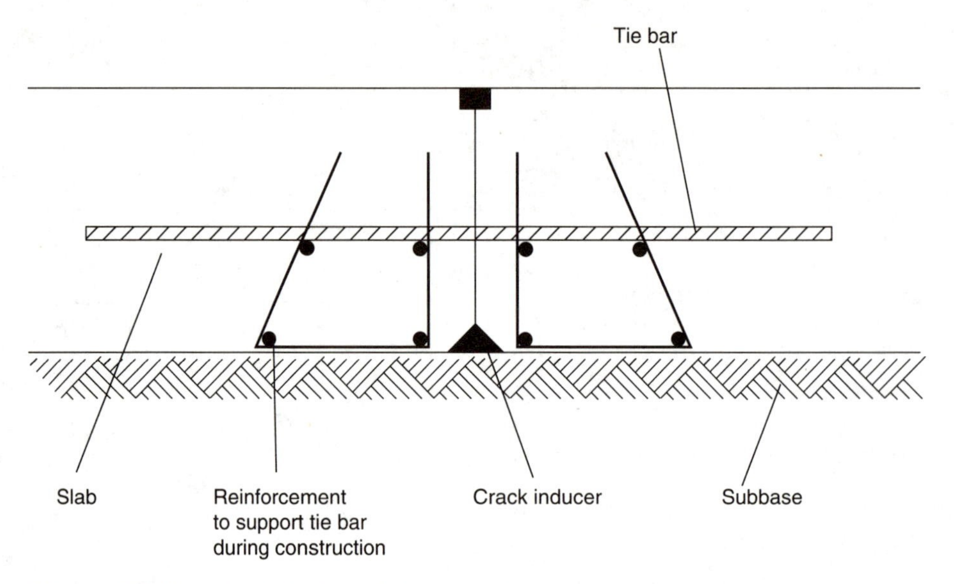

Figure 5.5 Warping joint for plain concrete

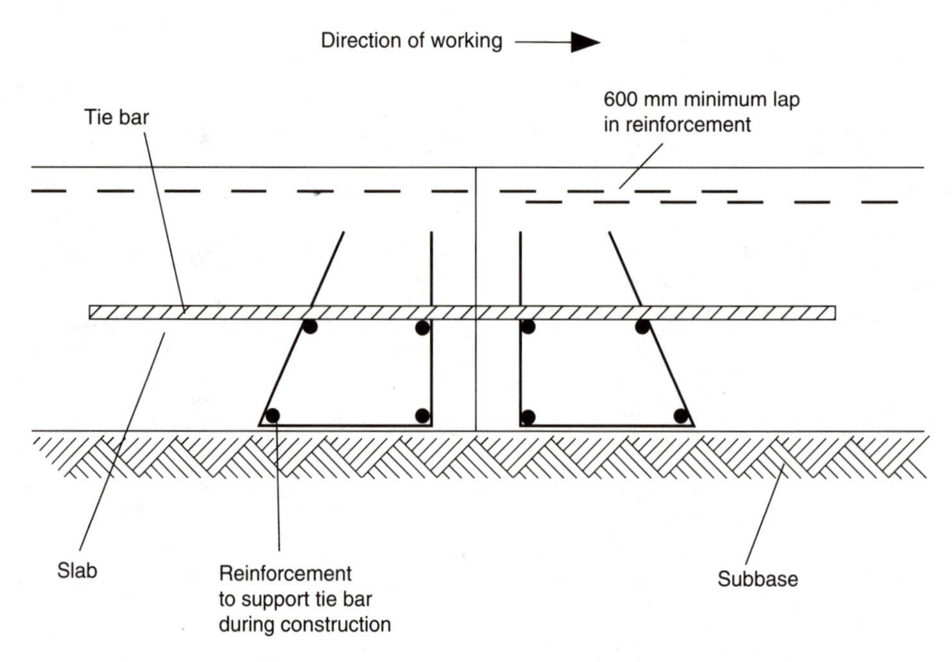

Figure 5.6 Construction joint

5.2.3.3 Longitudinal joints

Where reinforcement is provided in a concrete pavement it is usually aligned longitudinally since this is the direction of the greatest stresses. Transverse reinforcement is generally nominal only, its main function being to locate the longitudinal reinforcement at the required spacing. There is nevertheless a possibility of warping about a longitudinal axis and the provision of one or more longitudinal warping joints is worthwhile.

Where it is convenient to form the full width of the pavement at one pass, longitudinal joints may be similar to the warping joint shown in Fig. 5.5. Often, however, work is arranged so that the pavement is formed in two or more parallel strips at different times, in which case the type of construction indicated in Fig. 5.7 is more appropriate.

Typical joint dimensions are indicated in Table 5.2.

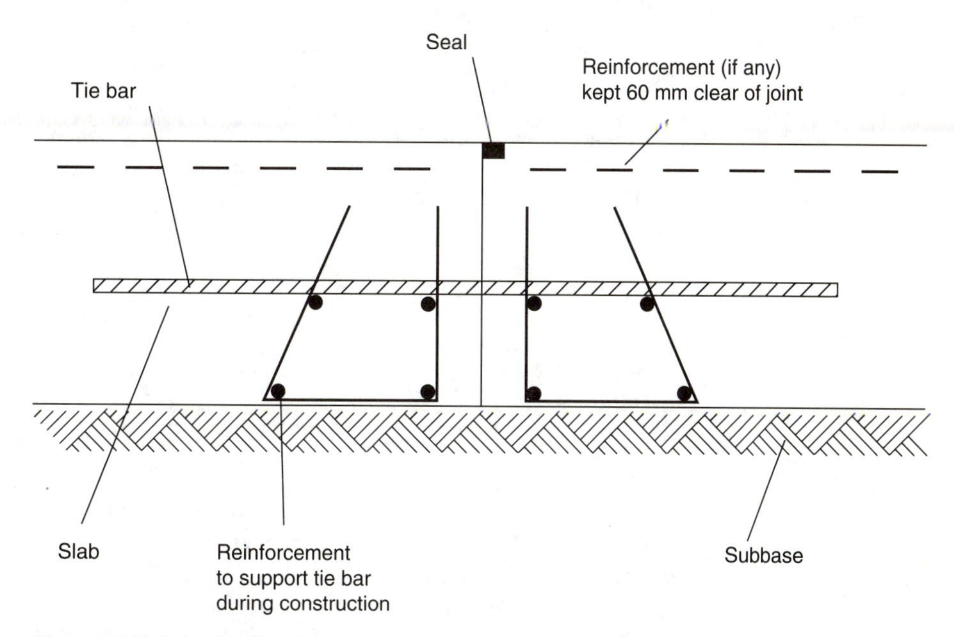

Figure 5.7 Longitudinal joint

5.2.3.4 Joint layout

The layout of joints in a jointed concrete pavement is an important determinant of its performance. For the joints to have the desired effect it is important that they should be positioned correctly, having regard not only to their correct spacing but also to the shape form of the slabs which are formed between them.

Transverse joints should be spaced so that the stresses developed in each slab do not exceed the strength of the concrete, including the effect of any reinforcement. Table 5.2 indicates typical design requirements.

Longitudinal joints should be provided so that slabs are not more than 4.2 metres wide for plain construction, or 6.0 metres wide where transverse reinforcement is provided – as is the case in the standard 100 × 400 mm 'long mesh' assumed throughout this section. Again these figures may be increased by about 20 per cent if the aggregate is wholly limestone.

Note that it is not good practice to arrange longitudinal joints so that slabs less than one metre wide are formed; these are prone to premature transverse cracking. Nor should the work be arranged so that longitudinal joints are formed in the wheel tracks.

Table 5.2 Jointed concrete pavements – dimensions of transverse joints

	Slab depth (mm)			
	150	*200*	*250*	*300*
Contraction joints				
Dowel bars	20 × 550	20 × 550	25 × 650	25 × 650
Dowel bar spacing	300			
Width of seal	0.001 × joint spacing, rounded up to the next 5 mm			
Depth of seal				
Hot poured	20–25	20–25	25–30	25–30
Cold poured	8–10	10–15	15–20	20–25
Expansion joints				
Dowel bars	25 × 650	25 × 650	32 × 750	32 × 750
Dowel bar spacing	300			
Width of seal	Filler thickness 25 mm. Seal 30 mm.			
Depth of seal				
Hot poured	25–30			
Cold poured	20–25			
Warping joints				
Tie bars	12 × 1400			
Tie bar spacing	400	300	230	180
Width of seal	10			
Depth of seal	15–20			

The notion of avoiding physically weak configurations or those which are likely to cause stress concentrations that the slabs will be unable to support is fundamental. Figure 5.8 illustrates this and shows various ways of arranging the joints at right angled and acute junctions. Figure 5.9 indicates typical treatments at roundabouts. Note in these details that transverse joints are always arranged to meet the edge of the pavement at right angles.

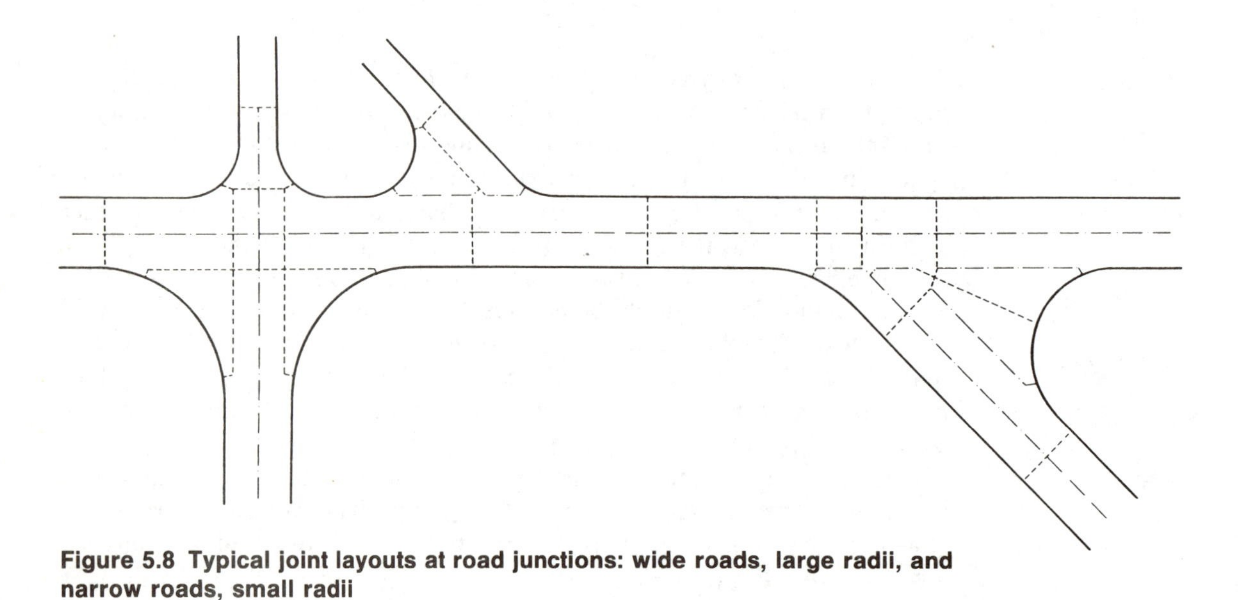

Figure 5.8 Typical joint layouts at road junctions: wide roads, large radii, and narrow roads, small radii

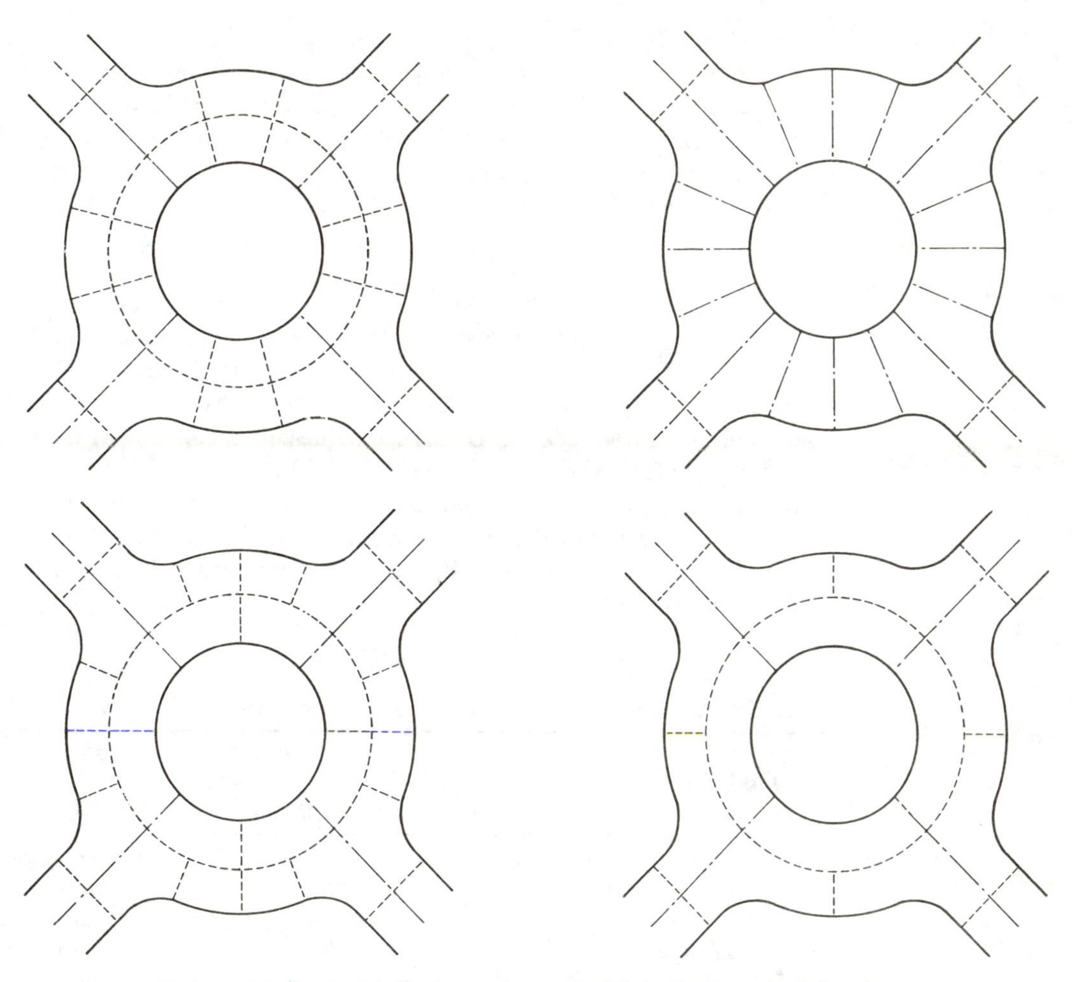

Figure 5.9 Typical joint layouts at road junctions: roundabouts

Transverse joints intended to allow movement should be continuous not only through the slab but also through any adjacent element likely to restrict or be damaged by such movement. This applies to kerbs and the like which may be bedded on the slab, and to structural elements supporting the pavement. In the common case where no such provision is made the designer should make sure that the slab can move independently of such external restraints; thus a separation membrane is generally provided beneath the slab to reduce frictional forces between slab and subbase.

Ironwork projecting through the slab merits special treatment. Access covers and the like are intended to be securely attached to their underground chambers which are of course immovable and inflexible, whereas movement is possible in the slab.

To provide the necessary isolation, small independent slabs are formed around such covers after the main slab has been produced. This is known as 'boxing out'. Butt joints similar to the expansion joint shown in Fig. 5.4 but with the dowel bars omitted – to allow relative movement and to facilitate excavation to the chamber below – are provided between the two slabs.

Additional diagonal reinforcement is needed in the main slab at the corners of the box to prevent cracks spreading into the main slab. The general bay layout and the position of ironwork should not allow boxes to be so close to each other or to the edges of the main work as to encourage the forming of cracks. Ideally boxes should be positioned at the ends of bays in order to reduce the length of joint.

5.2.4 Slab design – continuously reinforced concrete

Table 5.3 shows that the spacing of transverse joints in a JCP should be directly related to the amount of reinforcement in the slab. Increasing the amount of reinforcement increases the slab's ability to withstand the various stresses tending to cause cracking. Clearly, if sufficient reinforcement is provided the slab will be fully able to withstand these stresses and no transverse joints will be necessary. Such a slab is said to be continuously reinforced, and a pavement in which a continuously reinforced concrete slab forms the running course and the major structural element is known as a continuously reinforced concrete pavement (CRCP).

Table 5.3 Maximum spacing of transverse joints

Amount of longitudinal reinforcement (mm^2/m)	Joint spacing (metres)		
	Contraction	Expansion	Warping
0 Slabs 225 mm thick or more	6	60	In place of some contraction joints. See note (2)
Slabs less than 225 mm thick	4	40	
385	18	In place of every third contraction joint; this replacement may not be needed in summer	Not required
503	24	As above	Not required
636	30	As above	Not required

Note
(1) The joint spacing indicated may be increased by 20% if the coarse aggregate is wholly limestone.
(2) In plain concrete slabs, not more than three consecutive contraction joints may be so replaced with warping joints.

CRCP has the twin advantages of long life with little need for maintenance, and the ability to function well over ground which offers varying support. Providing that it is unlikely to be opened, perhaps to gain access to utilities underneath, the low maintenance needs make it particularly attractive where repair works are likely to disrupt traffic severely, and this to some extent offsets the initial cost premium. CRCP is also particularly suited, by virtue of the lack of transverse joints, to accept an overlay, thus simplifying maintenance when this eventually becomes necessary. Indeed, a major use of continuously reinforced concrete is as a roadbase in new construction, together with a bituminous surfacing. When used in this way the concrete element is known as a continuously reinforced concrete base (CRCB).

5.2.4.1 The design of CRCP

Whereas in the design of a jointed concrete pavement the intention is to minimise the stresses in the pavement by providing frequent stress releases and allowing the slabs some freedom of movement between these, in the case of a CRCP the intention is to ensure that no such stress releases develop and that the concrete slab is therefore kept at a uniform, high level of stress. Consequences of this altered approach include the absence of transverse joints, an increased provision of longitudinal reinforcement, and altered construction details at the ends of the slab, beneath it and at end-of-day joints.

Cracks usually form in CRCP. These need not give rise to concern, providing they are narrow and at a spacing of the order of 1.5–2.5 metres. If cracks form at a closer spacing than this, the possibility of failure of the pavement by punching out becomes significant; cracks formed at greater spacing are likely to be wider than 0.5 mm and therefore demand some maintenance. Vetter[3] has shown the average crack spacing to be a function of the tensile strengths of the concrete and the steel reinforcement, their elastic moduli, the proportion of reinforcement present in the slab, and factors related to the bond stress between reinforcement and concrete. The minimum percentage by volume of steel reinforcement required to prevent yield in the steel is such that

$$p_s = 100 \times f_c/(f_y - mf_c) \qquad [5.3]$$

or

$$p_s = 100 \times f_c/(2(f_c - tc_e E_s)) \qquad [5.4]$$

whichever is the greater; in which

p_s is the required minimum percentage of steel reinforcement
f_c is the tensile strength of the concrete
f_y is the yield strength of the steel
m is the modular ratio, equal to E_s/E_c
E_s is the modulus of elasticity of steel
E_c is the modulus of elasticity of concrete
t is the variation in temperature of the slab during its life
c_e is the thermal coefficient of expansion of steel
and the coefficient of friction beneath the slab is 1.5

The appropriate minimum proportion of steel to use in any particular application therefore depends on these variables, which are not necessarily standard between specifications. In the UK a typical proportion of high yield steel reinforcement is 0.60 per cent.

There is no generally accepted method for designing the slab thickness. Hitherto, typical approaches have been to design the slab as it if were jointed and then to reduce its thickness somewhat to allow for the increased life which may be expected from a CRCP. This reduction has either been a standard dimension of up to 50 mm, or a proportion of the JCP slab thickness of between 10 and 17 per cent. Such an approach is not wholly satisfactory, but arises from the imperfect understanding of the behaviour of CRCP.

In the analysis which led to equation [5.2], it was found that joint spacing was not a significant variable in those pavements studied. Major determinants of pavement life were the thickness and strength of the concrete, the amount of high yield steel reinforcement, and the support conditions beneath the slab. Clear parallels can be drawn between equation [5.2], intended to relate to jointed concrete pavements, and the case in hand. If we take the 28 day mean compressive strength of the concrete to be 48.2 MPa, as before, and the proportion of high yield steel reinforcement present to be 0.60 per cent then equation [5.2] gives us results which are consistent with the traditional approaches mentioned above, in all but the most heavily trafficked (more than 300 msa) situations. The results of such an analysis are shown in Fig. 5.10.

The position of the reinforcement is of interest. In JCPs reinforcement is placed at a depth of about 60 mm below the surface in order to control the opening of cracks. In CRCP we are also concerned that the cracks should be spaced with some uniformity and the reinforcement is sometimes placed at the centre of the slab for this reason. It has been mentioned that the reinforcement used should be high yield steel; the use of mild steel will significantly increase the amount required and hence the cost. In order that adequate bond should develop between reinforcement and concrete, deformed bars should be used. The standard long mesh commonly used for JCP is not suitable for CRCP, because the main bars are not large enough; alternative arrangements should therefore be made. It is often possible to have a designed mesh fabricated to specific requirements, which can not only simplify construction but also ensure that the amount of reinforcement accurately meets the design requirements. Such purpose made fabrics will of necessity be much heavier than those used in jointed pavements and will therefore demand stronger and more frequent supports. A typical arrangement is the use of ring spacers, two long bars held the appropriate distance apart by short vertical bars, the whole welded assembly being joined end to end to form a ring about 2 metres in diameter, which rests on the pavement foundation. Laps between bars should permit the full transfer of stress; a typical requirement for deformed high yield bars is that the lap length should be 50 times the bar diameter. Laps should be staggered so as not to create planes of weakness in the work.

Construction joints will be necessary since the whole of the work cannot usually be completed in one day or in one pour. These should provide full lap length. Another common design requirement is that the main reinforcement should be doubled for at least 750 mm on either side of the joint. Construction joints should not be continuous across the full width of the pavement but should be staggered to promote further stress transfer via the tie bars in longitudinal joints. It is bad practice to form 'leave-outs' – temporary gaps in the work – although these may appear to be expedient for associated works

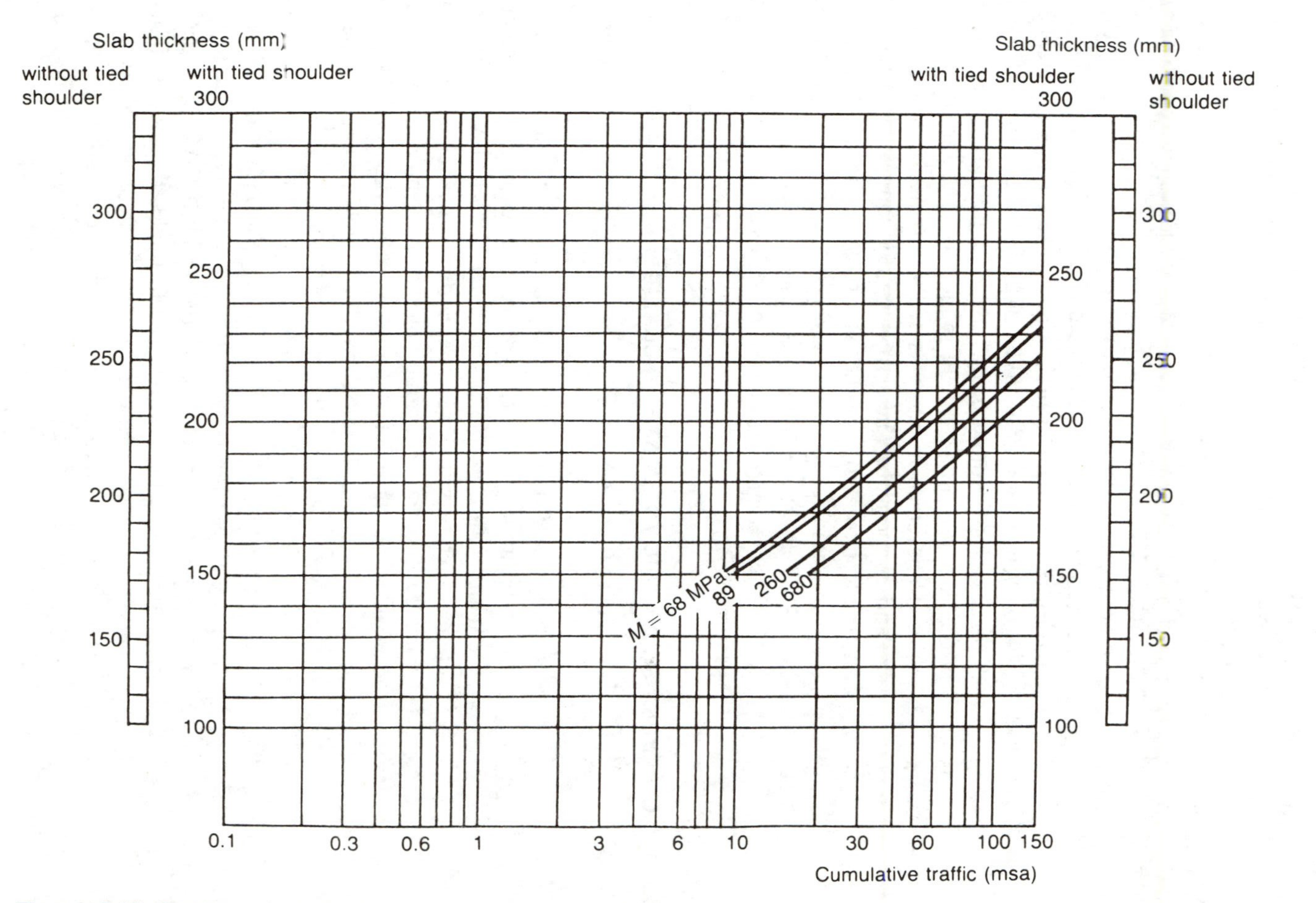

Figure 5.10 Continuous concrete pavement design chart: *R*=0.6 per cent

such as drainage or highway support structures; failure of the joints at one or both ends of the leave-out is not unknown.

In the body of an area of CRCP movement of the concrete is restrained not only by the presence of the reinforcement, but also by friction beneath the slab. For this reason it is desirable to omit the separation membrane generally included in a jointed pavement. Where the top of the pavement foundation is porous and grout loss may significantly weaken the slab, an increase in thickness of perhaps 20–25 mm is made to allow for this.

A further restraint to movement at any cross-section through the continuous slab is provided by the adjoining concrete. Towards the ends of the work, this may not be sufficient to stop movement and the cumulative effect of this can be a large amount of shrinkage and thermal movement at the ends of a length of CRCP. Provided that the length of pavement exceeds a certain minimum, the amount of movement is independent of the length of the pavement but can nevertheless amount to several tens of millimetres if the ends of the slab are unrestrained. Because of the difficulty of satisfactorily accommodating this amount of movement it is the practice to confine the ends of the slab so that such movement does not take place. Terminal anchors usually take the form of three or four ground beams projecting about 1.5 metres below the bottom of the slab into the pavement foundation and the subgrade. These beams are placed transversely at intervals of about 6 metres, starting about 2 metres from the end of the slab. The slab is rigidly connected to these by reinforcement passing from one to the other and by shear keys in the tops of the beams. An expansion joint of the type shown in Fig. 5.4 is provided between the end of the CRCP and the adjacent pavement.

The requirements for the width of CRCP are the same as those for a jointed pavement.

5.2.4.2 Continuous concrete roadbases – composite pavements

An alternative to CRCP which offers similar advantages of long life and tolerance of uneven subgrade conditions – such as might be found on redevelopment sites – is offered by the use of continuous concrete as a roadbase, with a bituminous surfacing above. This composite pavement avoids many of the maintenance problems associated with a concrete running surface and is often cost effective for the larger contract, when a use of a paving train becomes feasible. It is the commonest use of pavement quality concrete today. The flexible surfacing reduces the range of temperatures likely to be found in the concrete slab, thus tending to reduce cracking, and excludes moisture, thereby removing one source of failure at cracks.

For these reasons the amount of reinforcement may be reduced from that suggested by equation [5.3]. During the 1950s and 1960s a number of UK highway authorities built continuous reinforced concrete roadbases (CRCRs) which typically contained of the order of 0.25 per cent high yield steel reinforcement, and these have performed satisfactorily. In earlier years, many roads were paved with continuous concrete which is very lightly reinforced, the intention being either to provide a running surface, or as a roadbase upon which wood block surfacing (see Chapter 6) might be laid. Such pavements have often achieved lives in excess of sixty years.

There are currently two approaches to the design of continuous concrete roadbases. One is to provide a CRCR as mentioned above, in conjunction with 100 mm of bituminous surfacing. The general construction details are similar to those already described for CRCPs, with a reduced amount of reinforcement. Typically, 0.4 per cent high yield steel is provided.

An alternative philosophy has been adopted by some highway authorities where the approach has been to use a standard design whose life has been found to be 'long'. A typical standard design is 100 mm of rolled asphalt surfacing on 300 mm unreinforced pavement quality concrete on a suitable pavement foundation; no attempt is made to relate the design to the volume of traffic expected to use the pavement. This approach has, for example, been used on major traffic routes in London and has performed adequately for at least twenty-five years.

Terminal anchors are not generally provided in conjunction with CRCR. The flexible surfacing provides insulation against extremes of temperature.

5.3 Construction of concrete pavements

There are a number of objectives to be satisfied in the construction of a concrete pavement. These include:

- the location of the reinforcement in the required position
- correctly forming the joints
- forming the slabs to the correct width, depth, alignment and surface regularity
- aftercare.

On small jobs, the work is carried out almost entirely by hand and this is generally the most familiar way of working. Manual techniques are not suited to the larger project, however, where the use of specialised plant to perform essentially the same function is economically worthwhile. Hand construction involves the following work.

5.3.1 Location of reinforcement

The placing of concrete is a vigorous process which requires that the reinforcement, where present, should be firmly located in the correct position. Sheets of mesh should therefore be well tied together with binding wire to ensure uniform coverage of the whole pavement, and should be located vertically by one of two means. Where the concrete is to be placed in one pass, suitable 'chairs' may be made by bending reinforcing bars to allow secure connection to the reinforcement above and to prevent penetration of the pavement foundation and any separation membrane below. The necessary longitudinal laps – 40 bar diameters in jointed work and 50 in continuous – should be provided. The whole assembly of main and transverse reinforcement and chairs should be sufficiently robust to be capable of supporting several operatives engaged in concreting.

Alternatively, if the concrete is to be placed and compacted in two layers – perhaps to provide an air-entrained upper layer, or for the very purpose of locating reinforcement – the reinforcement may be placed on the lower layer after compaction. This will be satisfactory if adequate arrangements are made for subsequently compacting the upper layer.

5.3.2 Forming joints

In order that the joint designs illustrated in Figs 5.3–5.7 should be realised, thought must be given to their construction. Where the joints coincide with the position of formwork, dowel or tie bars can be arranged to project through the forms, and rebates for joint seals or to act as crack inducers can readily be included, for example by attaching these to previously hardened concrete before working away from such a joint. Where concreting is to be continuous past the position of a joint, typical arrangements include the fixing of bottom crack inducers to the pavement foundation, the supporting of dowel and tie bars either by their own system of chairs as shown in the figures or, where the adjacent slabs are reinforced, from the main reinforcement; and the cutting of surface chases in the green concrete at previously marked locations. Expansion joints, however, cannot be satisfactorily formed in this way.

5.3.3 Forming slabs

Each slab or group of slabs which is to be poured on the same occasion will be confined below by the pavement foundation, and at the sides and ends by formwork.

In most cases the top of the pavement foundation will be in an unbound permeable material into which it is quite possible for significant grout loss to take place from the freshly poured concrete, weakening the slab. To prevent this, and to reduce frictional forces, a separation membrane of heavy duty polythene or similar is provided between foundation and slab in jointed pavements. Alternative arrangements for continuous pavements are described above.

Formwork should be in purpose made road forms, typically of steel and made to the same depth as is required of the slab. Road forms are held in position by road pins, bars typically of 12 mm diameter which are driven through holes in the flanges of the road forms and into the pavement foundation beneath, to locate the forms in the correct position. The forms should also be set out to the correct vertical alignment.

Once the forms have been placed and the reinforcement checked and approved, concreting may begin. The initial aim is to ensure that the concrete fully occupies the intended space, and is properly compacted around and into the reinforcement and formwork. The wet concrete is vibrated to achieve this, initially by vibrating pokers similar to those used in structural work and finally by tamping the concrete with a tamping board – a board slightly longer than the width of the work which is used to beat and compress the concrete. This effort is applied either manually or mechanically, and by using the tamping board to screed the concrete to the required surface profile by drawing it to and fro on the road forms. Tamping will leave a coarse textured surface. If this is not required the surface of the wet concrete should be lightly swept by a soft broom to remove the most excessive of these marks and also to remove any surface laitance. The use of a wire broom will create a characteristic macrotexture which is often satisfactory as a running surface; the specification may require that the channels be left smooth. Brushing should of course be transverse.

It is also important that the surface of the work is properly regular and uniform. Typical specification requirements[6] are that the surface of a pavement should be within 6 mm above or below the design level, and that local irregularities should be no more than the maximum permitted for bituminous pavements and indicated in Table 4.9. Surface texture should be measured by the sand patch test – a smaller quantity of sand is used than is the case for texture depth measurements on bituminous surfaces – with the required minimum average of ten such tests being 0.65 mm, taken before the pavement is opened to traffic. Texture depth in excess of 1.3 mm has been found likely to cause excessive tyre noise.

Providing that normal precautions are taken, this should result in a slab whose physical properties are as expected. It is important that the concrete is protected from the weather while curing – a curing membrane on the surface will prevent dehydration and tentage may be used to provide protection from drying winds and the heat of the sun. Concrete pavements should not be trafficked by public or site vehicles for at least seven days after it has been placed.

5.3.4 Mechanised construction

On large jobs the manual construction process outlined above is not the most economic, and cost savings – and, often, higher quality – can be achieved by mechanised construction. In its general form this will consist of a series of machines which pass over the work and carry out the various activities already described. Such an arrangement of plant is known as a paving train.

Mechanised *in situ* concrete paving may be carried out either in a way directly analogous to the manual approach already described, using fixed road forms but with variations to take advantage of the potential benefits of mechanisation; or the work may be done without the use of fixed forms. This latter is known as slip-form paving and relies on the ability of the slip-form paving machine to spread and compact fresh concrete in such a way that the sides of the slab will be self-supporting before the concrete has developed its initial sct.

In either case, a paving train is so arranged that in one pass over the prepared pavement foundation the full construction process is completed, leaving a slab which may be trafficked after the usual curing period. The technology to achieve this is complex and somewhat specialised and is fully described elsewhere.[4,5]

The most satisfactory way to place reinforcement in fixed-form paving is to lay and compact the concrete in two courses, the reinforcement being placed manually on top of the first course after compaction and the second course immediately following. Two-course construction is often less successful where slip-form paving is used. Preferred alternatives include pre-positioning the reinforcement on chairs, as described for single course manual work, or by the use of pavers which are capable of accepting mesh reinforcement spread out in advance of the paving train, and incorporating it in the concrete slab.

Typically, a fixed-form paving train will consist of several items of plant, each with its own function:

- a feeder, to take delivery of the concrete as it arrives at the point of laying
- a spreader, to distribute the concrete across the width of the work (more than one lane may be produced in one pass)
- strike-off paddles to regulate the surface of the concrete
- compacters and finishers to initially stabilise the mix
- dowel and tie-bar placers to locate these elements in transverse and longitudinal joints
- joint groove forming and finishing equipment, where such grooves are not to be produced by sawing
- final finishing equipment, to further regulate and compact the concrete after the insertion of dowel and tie bars
- texturing equipment to produce the desired surface finish
- a curing compound sprayer, possibly followed by protective tentage.

These machines typically travel on the road forms whose tops are shaped to accept their flanged wheels. An arrangement suitable for two-layer working is shown in Fig. 5.11.

The slip-form paver produces a fully compacted slab and, in the absence of road forms, subsequent attempts at compaction are likely to be counter-productive since the unsupported edges of the fresh concrete will slump under the applied compactive loads. It is therefore not practical to place dowel or tie bars mechanically since the necessary disturbance of the surface of the work cannot properly be made good. Transverse joint assemblies are fitted to the pavement foundation or otherwise supported, as is the case in manual construction.

Slip-form paving plant is self-powered and automatically guided from a wire previously set up. The functions of spreading, compacting and finishing the concrete are all performed by the slip-form paver, leaving only the forming and finishing of transverse and longitudinal joint grooves, surface texturing and the spraying of curing compound to be done by following equipment. A typical arrangement is shown in Fig. 5.12.

5.4 Specification requirements

Material specification requirements for pavement quality concrete have been discussed in Chapter 2. To achieve these on site care must be taken that the design mix is consistently provided, that the concrete is properly compacted, and that adequate arrangements are made for curing – particularly during the first seven days.

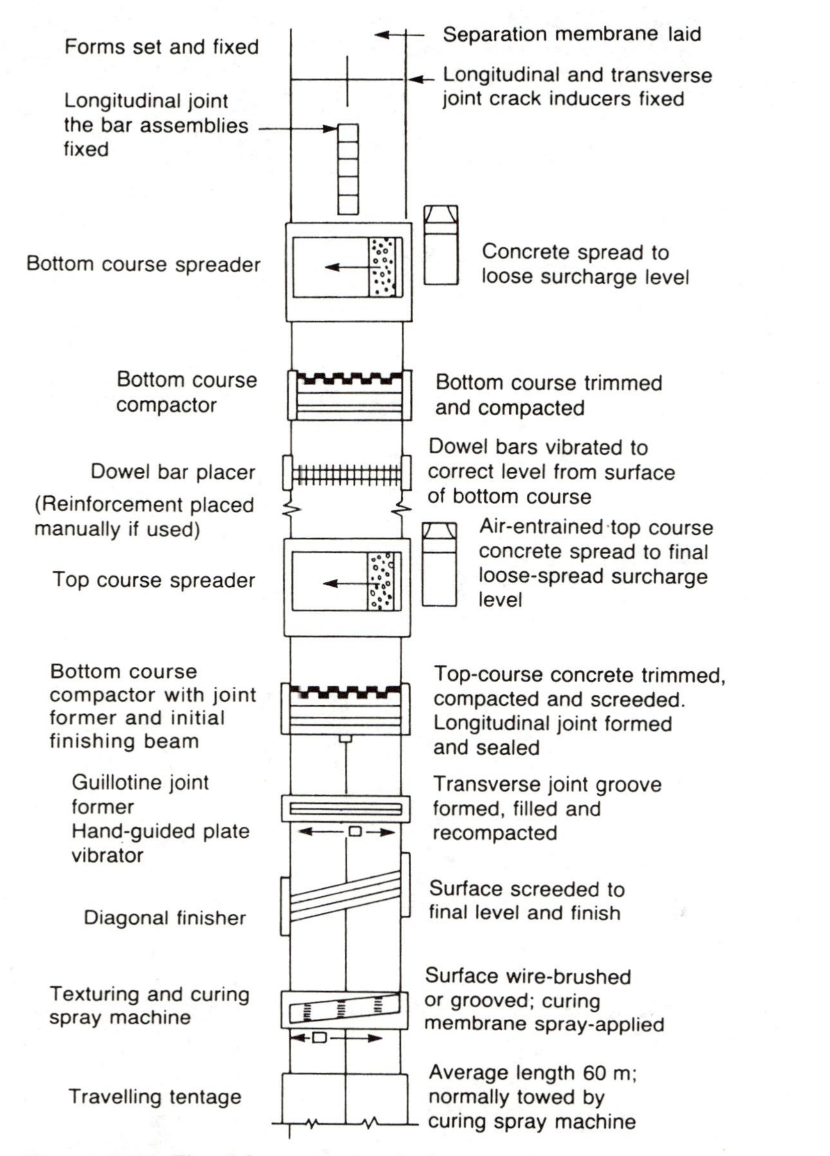

Figure 5.11 Fixed-form paving train

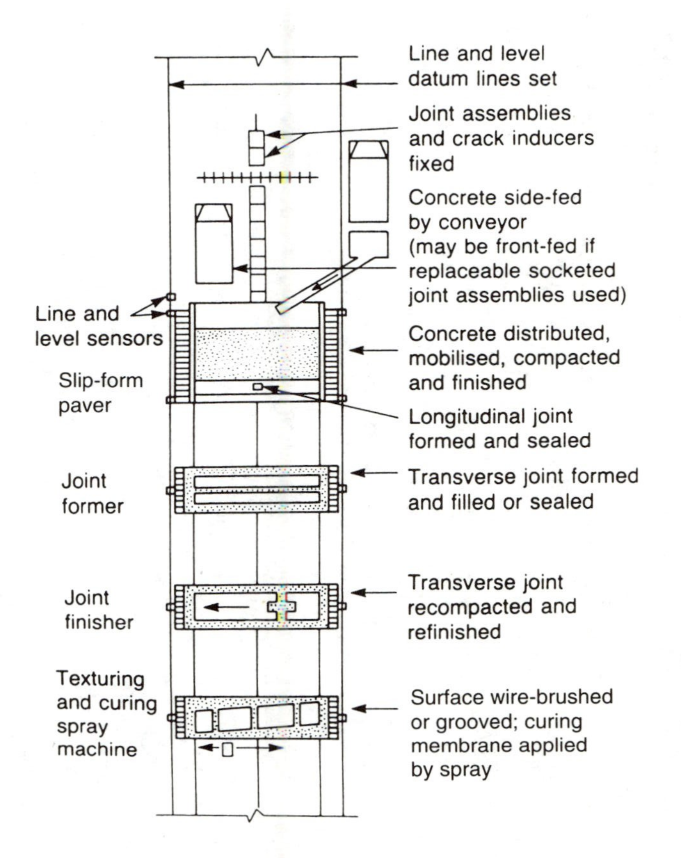

Figure 5.12 Slip-form paving train

Revision questions

1 Briefly describe two types of concrete roadbase. Compare these with bitumen macadam roadbase. In which circumstances would you consider each of these three roadbase types to be particularly appropriate? Why?

2 With the aid of sketches, describe and compare each of the following types of rigid pavement:
(**a**) unreinforced concrete
(**b**) jointed reinforced concrete
(**c**) continuously reinforced concrete.

3 Pavement quality (PQ) concrete has certain characteristics which necessitate the provision of joints, reinforcement or both when the material is used in the pavement. Describe these characteristics and explain the relationship between them and the design of joints and reinforcement, including the influence of any other physical factors.

4 The road shown in Fig. 5.13 is to be built as a jointed reinforced concrete pavement. It will be built in the UK during July and August. Subgrade CBR is 5 per cent; expected cumulative traffic is 5 msa. Prepare a design for the pavement, including a suitable joint layout.

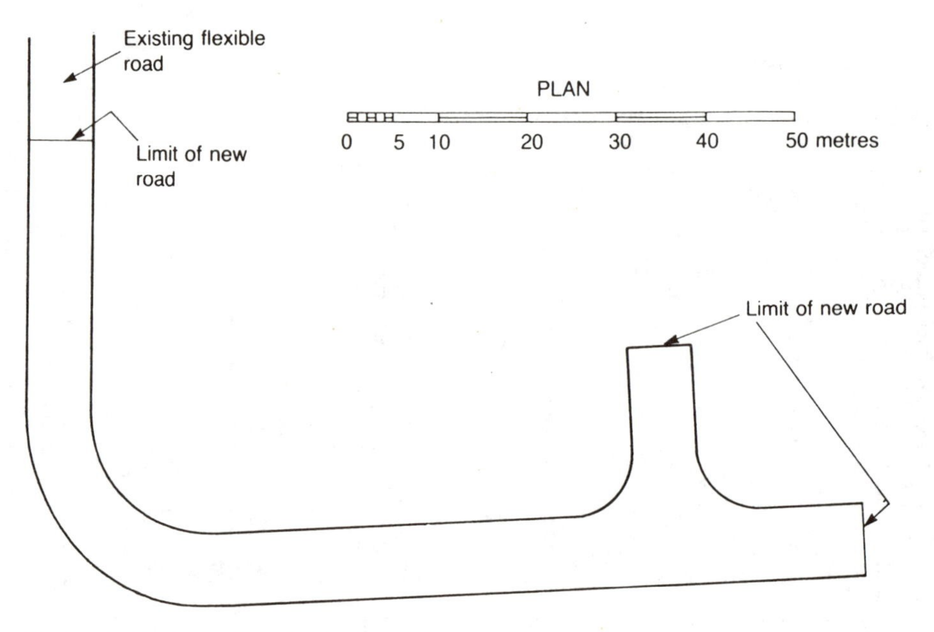

Figure 5.13 A small industrial service road

References and further reading

1 Westerggard, H. M., 'Stresses in concrete pavement computed by theoretical analysis', *Public Roads*, 1926, 7, No 2, pp. 25–35.
2 Mayhew, H. C. and Harding, H. M., 'Thickness design of concrete roads', TRL Research Report 87, Department of Transport, 1987.

3 Vetter, C. P., 'Stresses in reinforced concrete due to volume changes', *Trans. Am. Soc. Civ. Engrs*, 1933, No 98, pp. 1039–80.
4 Walker, B. J. and Beadle, D., 'Mechanised construction of concrete roads', Cement and Concrete Association, 1975.
5 'A guide to concrete road construction', DoT with C&CA, HMSO, 1978.
6 DTp Specification for Highway Works, *op. cot.*

6. Alternative paving types

The two preceding chapters have dealt with pavement types which are appropriate for major public roads. That is an important part of highway engineering, but there is more.

There is another significant area of highway engineering in smaller or less formal projects, including for example temporary roads for haul routes and other purposes, car parks, farm roads, cycle tracks, footways, mixed use areas to be shared by pedestrians and vehicles, pavements intended for abnormally heavy vehicles, minor estates roads, and so on. It is to the designer working in these areas that this chapter is addressed.

6.1 Temporary roads

Temporary roads are required in a variety of situations – for use as haul routes or temporary access routes in major construction projects, to facilitate military manoeuvres, to give access to temporary parking areas.

6.1.1 Unpaved roads

Haul routes have in the past often been constructed of no more than a layer of graded stone laid on the native soil. This arrangement has often suffered from the disadvantage that in wet weather the subgrade has been inadequately protected from moisture and rutting has occurred as shown in Fig. 6.1.

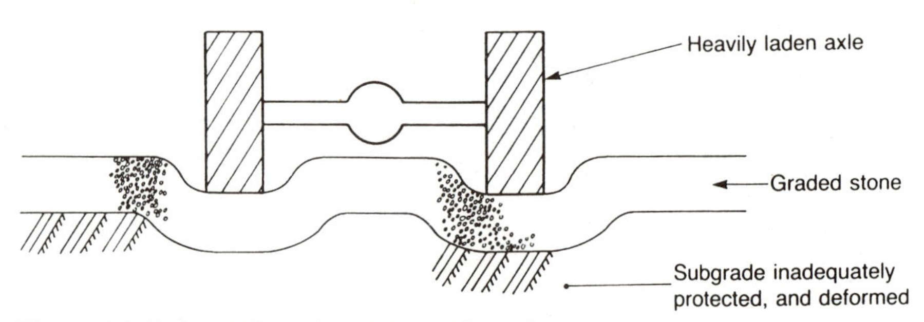

Figure 6.1 Deformation of an unpaved road

Rutting can be repaired by the provision of more roadstone, but this increases the cost. Alternatively, the situation can be improved by the inclusion of a geotextile or geogrid at the formation. In this application the membrane helps distribute the live loads transmitted through the unbound pavement to the subgrade (in contrast to the separation/filtration mode sought in a permanent pavement). Its ability to achieve this will depend on its strength, typically expressed as the load corresponding to a 10 per cent strain, and on the strength of the subgrade.

Current UK practice is to provide the depth of unbound material indicated in Fig. 6.2. The benefits of providing a suitable membrane beneath the granular material are most marked at low subgrade CBRs; a saving of 100 mm of imported material may be expected with CBRs in the 2–3 per cent range and traffic volumes as indicated in Fig. 6.2. If a geotextile is provided which can also act as a separation membrane there will be further benefits in that the internal drainage of the granular layer will be maintained, and the material may be salvaged for re-use elsewhere if required.

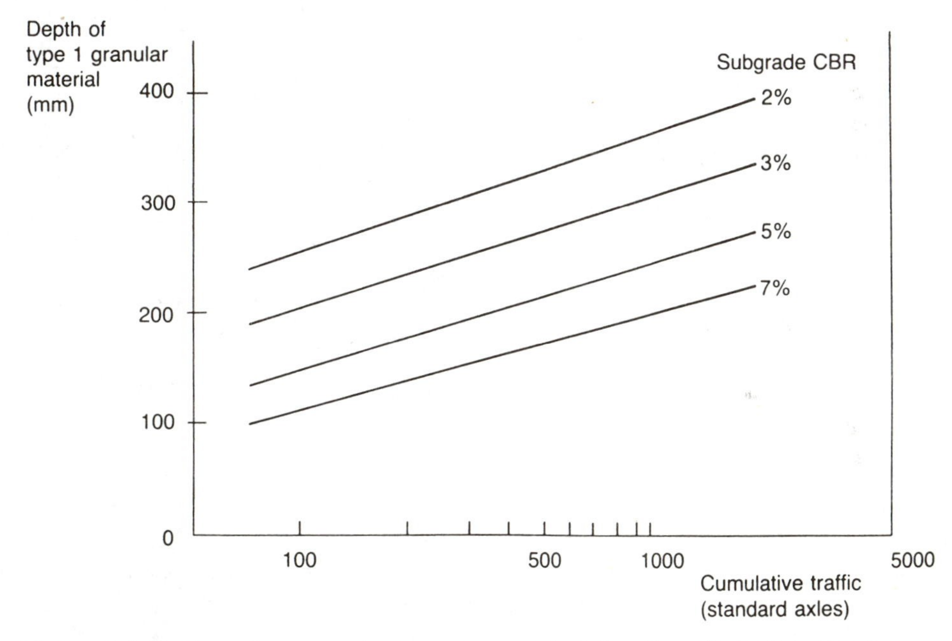

Figure 6.2 Design of unpaved temporary roads (after Powell, Potter, Mayhew and Nunn)

6.1.2 Other solutions

The prime function of a pavement is to spread traffic loads in such a way that the stresses acting on the soil are not excessive. In addition to the use of bound and unbound, graded and other stone generally discussed in this book, it is possible to achieve this objective by protecting the soil with a suitable mat or grid of structural elements.

In the past various materials have been used in this way. Corduroy roads were formed from timber poles, halved longitudinally and secured side by side

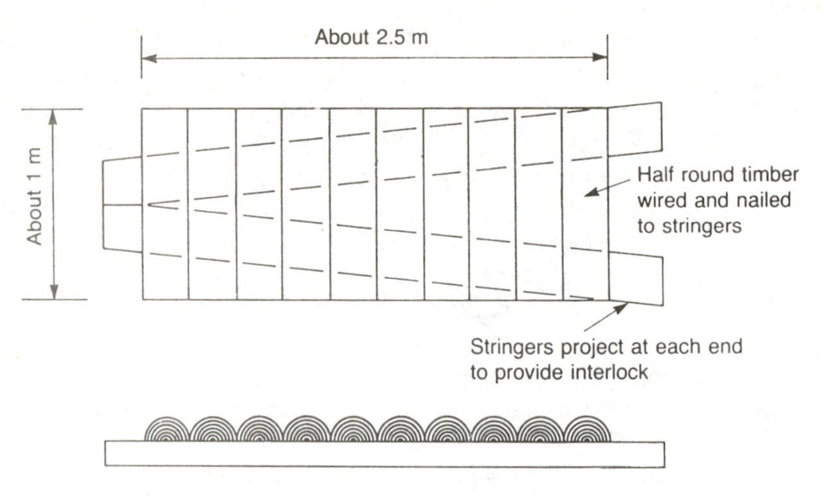

Figure 6.3 Corduroy road mat (plan and elevation)

to similar stringers to form mats which when laid end to end interlocked to give a passable route. This military expedient is illustrated in Fig. 6.3. Wire mesh and expanded metal have also been used in an attempt to spread the concentrated stresses caused by wheeled traffic.

A more sophisticated and more successful approach is offered by the trafficability mat, a geotextile woven from high strength polymer based materials with steel wire included in the warp and steel bar in the weft. Such a mat is supplied in rolls 4 to 5 metres wide and is simply unrolled over the surface, providing sufficient protection to allow many passes of commercial vehicles over ground of low CBR.

6.2 Light duty roads

These roads intended to be permanent features but which do not merit the heavier types of construction generally associated with public roads. Examples might include private drives, cycle paths, some car parks, farm tracks, and very minor rural roads in undeveloped areas.

Such pavements are very often not surfaced with an impermeable layer of material. In such cases the most important feature of the design is the provision of a generous camber, often in conjunction with ditches, to promote the free drainage of water from the surface of the road. An earth road represents the simplest such 'pavement' and consists of one or more 100 mm layers of a clay/sand/gravel mixture spread and compacted on a previously prepared formation on the native soil. A typical section through such a pavement is illustrated in Fig. 6.4(a). The ability of such a pavement to continue in service will depend on the volume of traffic to which it is subjected, and the climate. In suitably dry climates, such a pavement can be expected to support up to about 50 vehicle movements per day over long periods without showing severe signs of distress; in moister areas no such hopes can be entertained.

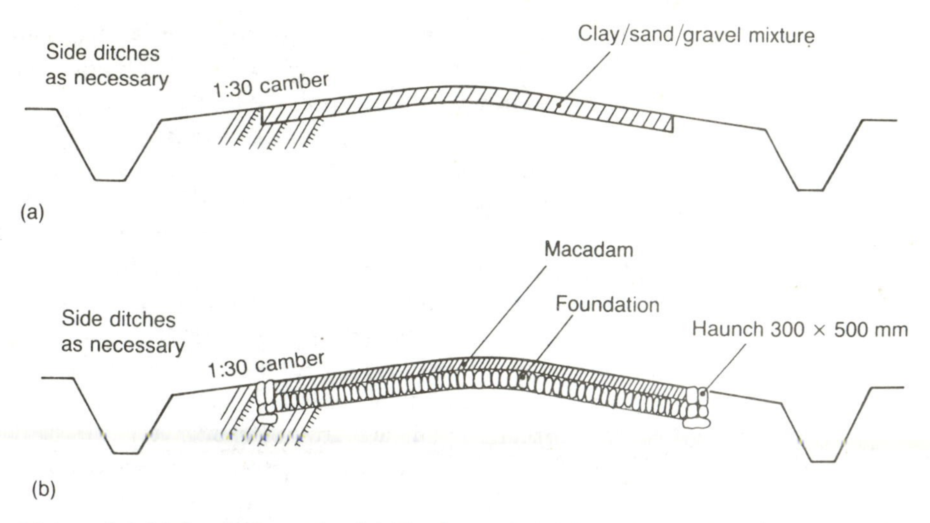

Figure 6.4 Light duty roads. (a) Earth road. (b) Waterbound macadam road

The logical development of this simple form is the waterbound macadam road. Once again, positive drainage is necessary, although in this case direct drainage of the formation is less important. The pavement is illustrated in Fig. 6.4(b) and consists of a well compacted formation on which is laid a foundation course of 250 mm stones or hardcore, the individual stones being laid on edge and across the road and bound together by the insertion of smaller stones of appropriate shape into the interstices. Lateral support to this layer is provided either by a kerb or by a haunch formed in similar large stones laid longitudinally at the road edge. The surface of the foundation is regulated by breaking off any large protuberences with a hammer, and rolled by a 10 tonne deadweight or equivalent roller.

The surface of a waterbound macadam road consists of a layer of what has come to be known as wet-mix macadam – that is a crushed rock or crushed slag graded in a specified way from 50 mm down to dust. This modern material is supplied premixed and should be laid at or about its optimum moisture content to form a subbase or roadbase in a modern pavement. Traditional waterbound macadam surfacing differed from this approach in that the coarse fraction of the material was laid and initially compacted, after which the fine fraction was spread on the surface, brushed in and rolled. The work would be sprinkled with water until saturated and rolled again and the process repeated, adding fines and water as necessary until a grout had been formed which would fill all the voids in the coarser material. The macadam would be left to dry out, and be opened to traffic.

Both earth and waterbound macadam roads suffer from the disadvantage that when trafficked by motor vehicles, dust becomes a nuisance. Not only is this unpleasant for travellers, it also involves a loss of material from the road which should eventually be replaced, creating a demand for almost continuous maintenance where flows of motor traffic are more than nominal. It was in response to this problem that the use of a bituminous binder was first introduced. We will consider here ways of providing such a layer at the top of the pavement in the context of a light duty road.

Historically, the response was to spray the surface with hot tar which cools to form an impermeable membrane and which binds the material at the surface into a coherent whole. The results obtained are, however, far from satisfactory in virtually all circumstances and so other approaches have been tried. The use of a bituminous emulsion mixed in with the aggregate in the appropriate proportions (about 7 per cent by weight of emulsion) and the whole then rolled, sealed with emulsion and blinded with fine chippings has often met with success. This is of course very labour intensive – and therefore often expensive. Alternatively the use of a minimal amount of premixed bituminous macadam is often worthwhile. A bound layer about 75 mm thick will withstand the damaging effect of motor vehicles of less than about 3 tonnes weight almost indefinitely provided that the lateral forces at the surface (due for example to sharply cornering vehicles) are not excessive.

Light duty pavements often experience their most demanding combinations of loads during construction. A car park built to a modest design may be capable of supporting cars almost for ever; but if the pavement foundation is incapable of supporting 15 tonne lorries bringing the surfacing material to the site, or the paving machine used to lay the material, than a satisfactory job will not be achieved. This design requirement may be satisfied by providing a pavement foundation as recommended by Table 3.1; by providing a reduced foundation and restricting the loads placed upon it during construction; or, less attractively but sometimes more expediently, accepting that some deformation of the foundation may take place during surfacing operations and relying on the bound layer to regulate the surface of the work to within tolerable limits. The approach adopted will depend on the desired quality of the finished job.

A typical minimal design for a car park on a subgrade of CBR between 3 and 5 per cent might be:

Bituminous grit blinding
75 mm of 28 mm nominal size dense basecourse macadam
150 mm granular subbase material
Single layer suitable geotextile or geogrid

The grit blinding will seal the surface against the ingress of moisture and will help to reduce the coarse surface texture of the material; the geotextile or geogrid will help to prevent failure of the foundation under the action of construction traffic.

It is not advisable to reduce the depth of the bound layer even further since the poor tensile properties of the materials available tend to result in early disintegration of the surface. A proprietary product is available which seeks to satisfy the need for good tensile strength characteristics while not demanding the use of such large quantities of material. This product is glass fibre reinforced surface dressing, formed by the application of a layer of suitable binder, a 'mat' of chopper fibrelgass strands blown onto the fluid binder, a further application of binder, and chippings spread and lightly rolled into the surface. This system provides a wear resistant, impermeable layer over any stable surface. The fibre reinforcement resists cracking of the layer, and, laid on an adequately compacted, reasonably dry and stable closed surface, is capable of providing a light duty pavement with very little foundation. Car parks consisting of little more than rolled earth, blinded hardcore and fibre reinforced surface dressing, have lasted well under continuous use by cars and light vans.

Fibre reinforced surface dressing is also suited to use on more heavily trafficked areas – in conjunction with a suitable designed pavement!

In addition to the possible cost savings associated with these light duty pavements, the use of a surfacing material other than coated macadam, asphalt or *in situ* concrete is often seen to introduce a pleasing visual contrast. In rural or semi-rural settings in particular the use of an uncoated stone in the surfacing can help to give a road or drive the appearance of a traditional country lane which is often very much in keeping the informal geometry and mixed vehicular and pedestrian use associated with lightly trafficked routes. Fibre reinforced surface dressing and waterbound macadam can both achieve this visual effect, although the latter is not suitable for surfaces where motor traffic is expected to be more than minimal. A similar effect can be achieved by the application of conventional surface dressing (Chapter 9) with an uncoated aggregate to a normal flexible or rigid pavement.

In urban areas the creation of an arcadian street scene is often not an objective, but the designer may still want to provide a surface which is more interesting in its appearance by virtue of the use of various surfacing materials than is possible with the normal range of construction techniques. Largely for historical reasons the usual approach to the problem of improving the appearance of urban streets takes the form of using small paving elements.

6.3 Small element paving

The relatively heavy traffic associated with main routes in towns and the comparative ease of raising taxes to pay for street works there led, throughout man's urban history, to the development of paved surfaces for urban routes hundreds if not thousands of years before the use of comparable techniques for inter-urban roads became commonplace. Before the nineteenth century the provision of a durable surface capable of withstanding the inroads of iron-shod hooves and wheels could only be achieved by the use of natural stone, either in blocks or in small slabs. Perhaps because of this historical background the view is commonly held today that the use of small paving elements is appropriate for urban streets – particularly where the appearance of the pavement becomes a major factor in its design. Locations where this is often the case include mixed use areas for residential developments, predestrianisation schemes and footways which will be subjected to overrunning by heavy vehicles, crossing or parking on the footway.

Small element paving has evolved, from wholly natural materials such as cobbles to the modern paving block in concrete or clayware.

6.3.1 Cobble paving

Cobbles are by definition naturally occurring water worn stones of a size suitable for paving. A cobblestone pavement consists of such stones, between 125 and 250 mm deep and rather less in length, set individually in a sand bed about 150 mm thick. Stones are laid to a random bond, rammed into place and covered with a 10 mm layer of sand to work into the joints under the action of traffic.

Such a pavement is little more than a historical curiosity today as it provides neither a comfortable nor a stable surface under the action of modern motor traffic.

6.3.2 Sett paving

This is a development of cobble paving which by virtue of a number of improvements provides a more comfortable, stable surface.

Setts are generally of granite or quartzite, dressed into blocks varying in size from 100 mm wide, 100 mm deep, 125 mm long to 125 mm wide, 150 mm deep and 250 mm long. The dimensions of the setts are determined by functional considerations; the width of the sett corresponds to the size of a horse's hoof – the joints giving a better grip: the depth of 100 mm is the minimum required for stability – greater depths allowing the surface to be redressed when worn; and the length is such that the setts can be handled easily.

Sett paving is generally laid on a concrete foundation at least 150 mm thick. For lightly trafficked applications an unbound foundation of type 2 granular material has been found satisfactory. The depth clearly depends on the properties of the subgrade. Today setts are usually laid in a cement mortar bed and the joints are filled with the same, a joint width of 10–15 mm commonly being achieved. Such a surface lacks flexibility and sometimes cracks. To provide an element of flexibility it was formerly the practice to lay the setts on a 25 mm dry mortar bed, to fill the joints with 6–8 mm gravel or stone chips, and to stabilise the system by sealing the joints with a bituminous grout. In either case a modern disadvantage to sett paving introduced for its visual appeal is that the joints if not formed carefully can be both wide and irregular, spoiling the look of the work. It may be impossible to avoid this if the setts are merely squared by a sett hammer in their manufacture, but with 'nidged' setts, dressed on five sides by chisel, close joints are possible and a pleasing effect can be achieved.

To provide good drainage setts should be laid longitudinally in the channel to a width of about 300 mm. In the body of the road they should be laid transversely to a stretcher bond; the consequent close spacing of the transverse joints promotes good frictional properties. The herringbone bond recommended for block paving is often not appropriate for setts since they lack the necessary regularity of size. It is possible to lay setts to more interesting and intricate patterns. Typical arrangements include concentric circles, either individually to emphasise features in the street scene, or in sequence over the area to be paved; and circular arcs springing from the channel or other longitudinal features. This latter arrangement is illustrated schematically in Fig. 6.5. Setts of cuboidal form are best for this type of work, particularly towards the centres of circles where radii are small.

Figure 6.5 Sett paving laid to arcs. Note longitudinal courses of setts to promote free drainage along the channel

6.3.3. Wood block paving

This again is a historical curiosity but is still occasionally found in the body of old roads.

Sett paving suffered from the disadvantage that the passage of horse-drawn traffic was a very noisy affair as the steel-shod tyres and iron-shod hooves clattered over the uneven, hard surface. A response to this problem was often to replace the stone setts with wood blocks.

Blocks were generally of softwood, preferred for its superior cushioning effect, steeped in creosote and laid with the grain vertical. The unit size was typically 150–225 mm long, 75–125 mm wide and 80–125 mm deep. The work was laid with open joints about 2–3 mm wide into which a bituminous grout was poured. Grout was also applied to the surface, which was coated with a protective application of grit. Wood block paving made no significant contribution to the strength of the pavement.

With the increased use of the motor vehicle, wood block paving was found to be unacceptably slippery, particularly in the wet, and its use was discontinued.

6.3.4 Block paving

Block paving consists of a large number of paving blocks laid on their backs in a course one block deep, in a laying course of graded sand, the whole system being placed on a suitably prepared foundation.

The paving blocks themselves are made of concrete or of clayware and are of the approximate size 100 mm wide, 200 mm long and between 65 and 100 mm thick. They are essentially rectangular in shape, but may have either linear or profiled edges. If made in concrete it is a simple matter to add pigments to the mix to achieve blocks of any desired colour so that patterns may be formed in the work.

The appearance of block paving is often considered attractive. In addition, block paving can be a particularly useful paving material in that it provides a very strong and very flexible surface which can support concentrated loads for long periods without suffering deformation, and because no complex or expensive plant is involved in its construction. An understanding of the way in which block paving achieves its strength is important to the successful use of the material.

Since the blocks are set in a matrix of well compacted sand, and since the joints between the blocks are thin, frictional forces between the blocks are large. Furthermore, providing that the blocks are deep enough they will resist rotational forces caused by eccentric loading. Figure 6.6 illustrates this. A layer of block paving is thus able to withstand forces set up in it as part of a pavement providing that the sand in the joints remains properly compacted, and providing that the blocks are thick enough to avoid rotation. This condition is known as one of interlock.

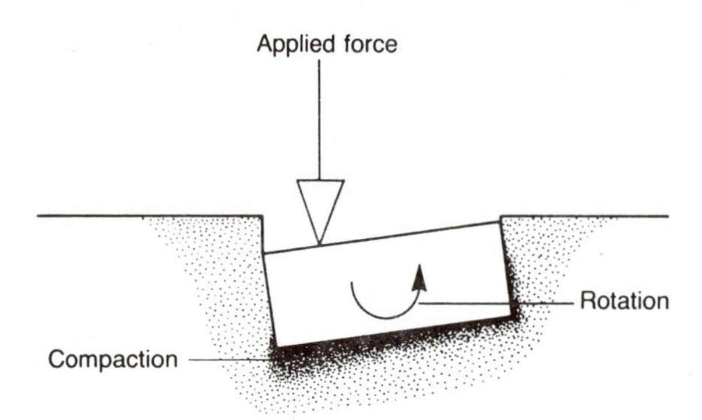

Figure 6.6 Effect of eccentric vertical force on a paving block

Often frictional forces are incapable of preventing all lateral movement and so the designer seeks to prevent the movement of individual blocks by laying them to a pattern or bond so that no block may move in isolation and so that no lines of weakness exist in the surface. Figure 6.7(a) indicates the arrangement known as parquet bond; while pleasing in appearance, this bond suffers from the disadvantage that it is criss-crossed by a series of continuous joints mutually at right angles. These continuous joints reduce the ability of the paving to resist lateral forces and it is for this reason that parquet bond is unsuitable for any surface which will carry vehicular traffic. Figure 6.7(b)

Figure 6.7 Block patterns. (a) Parquet bond. (b) Stretcher bond. (c) Herringbone bond

illustrates stretcher bond, an arrangement which offers some improvement in performance over parquet bond but which still suffers from the disadvantage of continuous joints crossing the work offering relatively little resistance to lateral deformation.

In Fig. 6.7(c) is shown the generally preferred arrangement for trafficked block pacing, known as herringbone bond. This bond avoids the formation of long joints in the work and thus provides a surface which is sufficiently stable to resist the lateral forces set up by the passage of vehicles.

As a refinement to the basic rectangular shape, and in an attempt to enable stretcher or parquet bond to be used in carriageway areas, some manufacturers have introduced shaped blocks whose sides are designed to interlock with one another, thus improving the lateral stability of the work.

Shaped blocks enable stretcher bond to be used successfully for roads and parking areas, provided that the continuous joints in the work are kept perpendicular to the direction of traffic flow; special radial shaped blocks are available to enable this arrangement to be maintained round bends.

The highway engineer must bear in mind the possibility that at some time in the future an opening may be made in the surface of the pavement to reach pipes or cables buried beneath. Reinstatement of such an opening will demand the supply of a small number of blocks to match any of the originals lost when making the opening. Each type of shaped block is generally made by one supplier only, is protected by patent, and requires the use of relatively complex moulds in its manufacture. If for any reason the original supply of shaped blocks were to cease it would thus be effectively impossible for an alternative matching supply to be found. Satisfactory resinstatements would thus become impossible unless a stockpile of blocks was held by the maintenance organisation. For this reason many highway authorities are now reluctant to accept shaped blocks in areas where openings may be made.

6.3.4.1 Characteristics of the surface

Block paving consists of a large number of small, rigid slabs with flexible joints between. The nature of a block paved surface may therefore depend on the blocks, the joints, and the way in which the two relate – as well as the foundation on which the blocks rest.

The texture of the surface is determined by the texture of the blocks. These are generally formed smooth; there is no macrotexture other than that provided by the joints. Consequently, drainage of the surface is not immediate. A film of water covers the contact area between wheel and road, and skidding resistance relies upon the microtexture of the aggregate exposed at the surface. This microtexture is often questionable in the case of clay blocks.

The joints are filled with sand and make no direct contribution to the performance of the composite material as a road surface. The ability of the joints to accommodate large deformations has been noted.

Apart from these considerations, the form of the work as a whole affects the passage of traffic in that the frequency of joints and the open nature of these provides a very harsh ride when trafficked at speed. The surface requires special care when roadmarks are to be applied in paint of thermoplastic. The careless reinstatement of areas where alphanumeric roadmarks have been applied on block paving subsequently opened can result in curious anagrams!

6.3.4.2 Suitable applications

Block paving is suitable for sites where the traffic speed is unlikely to exceed about 50 kph (30 mph).

The appearance and nature of the finished surface is often considered to be attractive and as such block paving is commonly used where the human scale of the material is of value; in residential developments, or where large areas of paving are to be 'broken up' visually to relieve an otherwise monotonous street scene.

The ability of block paving to withstand concentrated loads has also led to its use at container terminals, bus stations and similar high load sites.

6.3.4.3 Block paving as an impermeable layer

Early in the life of the pavement, the joints are filled with sand and are therefore permeable. As the surface is trafficked, the laying course becomes fully compacted, interlock develops and fine surface detritus tends to seal the tops of the joints. The amount of water entering the pavement is then reduced.

When designing areas paved with *in situ* slabs or with bituminous surfacing the designer may reasonably expect the surface to be impermeable. This is not the case with block paving and allowance should be made for the entry of water from above into the structure of the pavement. Unbound granular materials within the pavement should therefore be such as will not be adversely affected by the presence of water. If necessary the subgrade may be protected by the inclusion at formation level of a thin impermeable membrane. The provision of positive drainage should be considered.

6.3.4.4 The design of block paving

In this context, block paving may be viewed as a very strong flexible surfacing. It may be assumed that a layer of 80 mm blocks together with the accompanying laying course has a similar load-spreading ability to that of about 160 mm of rolled asphalt. Current design practice is based upon the approximation, which is probably conservative. Figure 6.8 gives recommendations for roadbase thickness, in conjunction with 80 mm block paving on a 50 mm nominal thickness laying course and a subbase consisting of 225 mm of type 1 granular material, on a subgrade of effective CBR of 5 per cent.

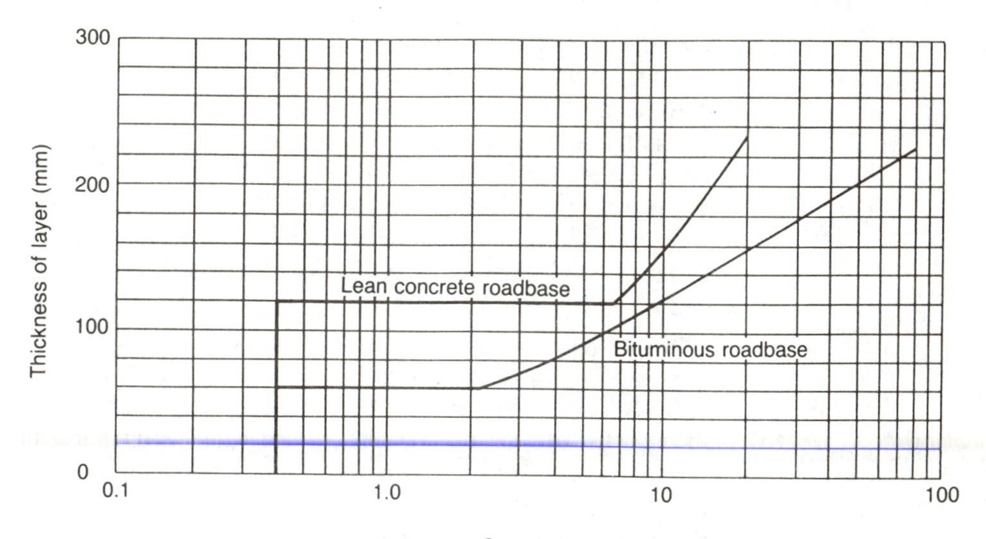

Figure 6.8 Thickness of roadbase in conjunction with 80 mm paving blocks, 225 mm granular subbase and 5 per cent subgrade CBR

The design requirements for the roadbase and subbase materials have been discussed in previous chapters.

It is also necessary to consider the thickness of the blocks, their colour, shape and bond, and edge restraint and construction details.

(a) *Thickness*

In order to prevent rotation, it is important to ensure that an adequate block thickness is provided.

Three standard thicknesses are generally available – 65, 80 and 100 mm. Of these, the thinnest is suitable only for footway loading and for light traffic conditions such as areas only used by cars and light vans. 80 mm blocks are capable of supporting all normal traffic with pneumatic or solid rubber tyres but for particularly punishing applications such as where small heavily laden steel wheels will be placed on the surface in conjunction with the absence of a roadbase the increased stability provided by 100 mm blocks may be desirable.

(b) *Colour, shape and bond*

Block shapes and bond have already been described with regard to their importance in providing lateral stability.

Concrete block paving is an unusual material for the highway engineer in that a number of colours are available – a typical colour range from one supplier might include natural (grey), charcoal (dark grey), and muted shades of red, yellow and brown. By careful selection of block colours the designer can ensure that the finished pavement harmonises or contrasts with its surroundings as desired, and also that informal roadmarks are permanently provided. It is also possible to bond thermoplastic road paint to blocks, but only in the workshop. 'Precoated' blocks can be used to provide yellow lines to indicate waiting restrictions.

(c) *Edge restraint and construction details*
The provision of proper edge restraint is vital to ensure that the laying course material in which the blocks are laid remains fully compacted. This is generally straightforward but difficulties can arise with some older types of ironwork where flanges and webs are present on the outside of the frame. These details can disrupt the laying of the blocks, a problem which can be overcome either by replacing the offending ironwork with purpose-built units which have no flanges, or by bedding specially cut blocks on a cementitious or epoxy mortar bed on and around the original flange and web. The forming of an *in situ* concrete collar around such ironwork is not only unsightly but also often unsuccessful; cracks often develop in the collar, leading to its eventual failure.

A second source of difficulty with ironwork, particularly at gullies where the presence of large amounts of surface water tends to exacerbate the problem, is the loss of laying course material through fissures in the brickwork into the chambers beneath. Sand loss here cannot easily be detected early in the pavement's life since the consequent accumulation of sand in the chamber cannot readily be distinguished from that washed from the surface of the work prior to the development of stable conditions in the tops of the joints. The fault therefore first becomes apparent with the onset of movement at the surface. A solution which has been found effective is to lay a sheet of geotextile under the laying course and up the side of the frame, to the surface. The fabric does not interfere with the laying process and is readily trimmed with a blowtorch to suit the finished surface profile.

6.3.4.5 Laying block paving

The essential features of the construction process are:

(a) the laying course is prepared to receive the blocks
(b) blocks are placed on the uncompacted laying course, cut as necessary to give a close fit at the edge restraints
(c) the paving is then agitated by a vibrating plate compactor, the effect being to cause the laying course material to flow into the open joints and to be compacted, thus providing conditions in which full interlock can develop
(d) opening to traffic and aftercare.

(a) *Laying course*
The functions of the laying course are twofold: to fill any irregularities in the surface upon which the material is to be laid; and to bed the paving blocks and allow filling of the joints so that interlock may develop.

Laying course material complying with Table 6.1 is spread on a subbase or roadbase, which should have been formed to the appropriate tolerances – typically ±20 mm and ±15 mm respectively. In such cases the laying course is best spread in two layers, the first of which is lightly compacted to a thickness of 35–40 mm in order to regulate the underlying surface and the second of which is laid and screeded but not compacted before placing the blocks, to a thickness such as will ensure a final thickness of about 50 mm in the compacted

Table 6.1 Grading requirements for laying course material

BS sieve size	*Percentage passing (by weight)*
5 mm	90–100
2.36 mm	75–100
1.18 mm	55–90
600 μm	35–59
300 μm	8–30
150 μm	0–10

laying course. Where a higher degree of uniformity has been achieved in the surface of the course beneath, the laying course may be reduced in thickness by up to one-third, and be laid in one screeded uncompacted layer prior to block laying.

(b) *Placing blocks*

Blocks are placed by hand, hand-tight, to the appropriate bond on the freshly laid undisturbed laying course. Work should proceed from an existing laying face or from an edge restraint, operatives positioning themselves on the new work and supplies of blocks being brought to them from the rear so as to avoid disturbing the laying course.

Full blocks should be laid first and part blocks cut as necessary to fill any gaps at the edges of the work. Blocks may be cut either by a simple block splitting machine or, less conveniently, by a bolster and club hammer.

(c) *Compaction*

Once an area of blocks has been placed it should be compacted with a small vibrating plate compactor. Too great a force will damage the faces of the blocks. The action of the vibrating plate is threefold: the sand beneath the blocks is compacted; the surface of the work is levelled; and the laying course material is agitated so that it tends to flow into the joints. The latter is particularly important in the development of interlock. Sometimes not enough sand flows up into the joints. Sand may then be added from the top by brushing over the surface, after which the vibrating plate is once again applied to the work. The sand applied from the surface must be fine textured – able to pass a 1.18 mm sieve – and dry, so it will readily flow into the joints. This process is known as top-filling.

(d) *Opening to traffic and aftercare*

Once surplus and has been swept from the surface, traffic may be admitted immediately.

Early in the life of a block paved surface, some sand loss from the joints is to be expected since the laying course material is unbound and open textured with the result that the action of the weather – particularly surface water – is quite effective in its removal. Where the surface is heavily trafficked and vehicle speeds approach the upper limit for which this material is suitable, the hydraulic action of the tyres in pumping surface water away from the interface between wheel and pavement can have a severe effect on the sand in the joints in prolonged wet weather. The loss of up to 20 mm of sand from the joints has been observed within a few weeks of the opening to traffic of areas of block paving.

It is clearly important that this process be arrested before interlock is lost. Appropriate treatment consists simply of once more top-filling the affected area, as described above. The treatment may in extreme circumstances need to be repeated; but the first period of dry weather will enable a surface seal of dust, rubber particles, oil and other road detritus to accumulate in the joints to form a more or less impermeable seal at the surface.

6.3.5 Thick paving slabs

For many years paving slabs for use in the footway have been made of concrete, precast to standard sizes about 600 × 600 × 50 mm and 900 × 600 × 50 mm. These have performed quite adequately for foot traffic, but where vehicles use the footway, legitimately or not, serious cracking of the slabs will occur. To overcome this difficulty some manufacturers now provide precast slabs smaller in plan and thicker than the old type; these thick paving slabs can withstand the action of traffic to a greater or lesser degree.

The original thick paving slabs were 450 mm square and 70 mm thick. These slabs can withstand occasional trafficking by commercial vehicles – as defined in Chapter 1 – and more frequent use by lighter vehicles. Other paving units have become available 400 mm square and 65 mm thick – dimensions which have the advantage of conforming with those of many paving blocks intended for use in lightly trafficked areas, and which give the slabs equivalent strength properties to the earlier design. Where the surface is likely to be used by commercial vehicles on a regular basis an increase in slab depth to 100 mm is generally appropriate since any irregularities in the support beneath the slabs can otherwise result in failure in such circumstances. Such a situation is not uncommon – a frequent use for these slabs is in the pedestrianisation of busy shopping streets, where delivery vehicles may often be obliged to continue to use the street after its closure to all other traffic.

Typically, thick paving slabs are laid on a sand bed overlying a subbase about 100–200 mm thick – depending on the soil conditions and the expected volume of traffic. Where this volume is more than nominal, a foundation and roadbase should be provided as for block paving. Note that this material is not suitable for surfaces intended primarily for use by motor traffic.

Thick paving slabs are available square and in hexagonal format, the latter intended to provide a more pleasing appearance. Many colours and surface textures are available, enabling interesting visual effects to be achieved. Variations in surface texture can be used in pedestrianisation schemes and elsewhere to provide tactile reference points for the blind – for example to delineate the end of the mixed use area and the start of the highway proper.

6.4 Urban design

Where service conditions permit, the imaginative use of the variety of surfacing materials described above can be of great help in improving the appearance of a project. Care should be taken in the formulation of such a design to ensure that not only are the paving techniques proposed appropriate to the adjacent buildings in terms of style and appearance, but also that the way in which these techniques are used in creating patterns or other arrangements in the surface are appropriate in both appearance and general character to the uses to which the surface will be put. The designer engaged on such a project is able to make a significant contribution in urban areas to the quality of the street scene and should ensure that the work is approached in an informed and professional way.

6.5 Kerbs

Kerbs are provided at the edge of the carriageway for several reasons:

- to define the edge of the carriageway and provide a degree of containment for traffic
- to form a channel as part of the surface water drainage system
- to contain the material of which the carriageway is made
- during the construction process, kerbs provide a continuous reference datum.

Kerbs are made of durable materials, set on a foundation which is strong enough to resist vehicle impact and the forces set up when adjacent granular material is being compacted.

On the larger contract it may be worthwhile to use the specialist plant needed to form kerbs *in situ* from concrete or asphalt. Usually that is not the case, and preformed kerbs in either concrete or natural stone – usually granite – are generally used. Granite kerbs are the more durable of the two and are often thought to look more attractive. However, their hardness makes them difficult to cut and dress, their weight can make them difficult to handle and their cost can be prohibitive.

Precast concrete kerbs are provided in the UK to conform with BS 7263 to various patterns and to several radii. Radius kerb is necessary where the kerb is to be laid to a radius of less than about 12 metres, to avoid unsightly gaps at the joints. Similarly, where it is required to form an external right angle in the kerb, quadrants are available to provide a neat appearance.

A specially profiled kerb is available which is capable of containing vehicles much more positively than the standard types illustrated in BS 7263. This 'safety kerb' requires a more robust foundation. It is designed to deflect the wheels of vehicles leaving the carriageway obliquely, containing traffic positively and without causing damage to the vehicle. Maintenance costs are much lower than for other types of vehicle containment barrier.

Small edging kerbs are available in concrete and sometimes in timber. The latter has a limited life. Edging kerbs are used at the back of footways.

All three types of kerb are provided in unit lengths of about 900 mm. An alternative is provided by some manufacturers in conjunction with block paving, in the form of small kerb units whose length matches the 100 mm or 200 mm module of block paving.

Construction details for precast concrete kerbs are shown in Fig. 6.9.

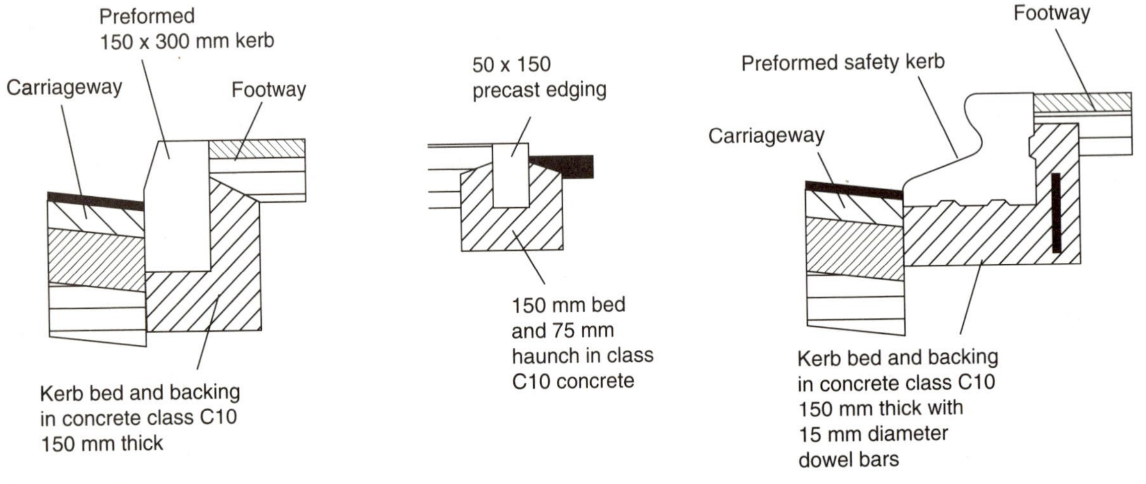

Figure 6.9 Kerb details

Revision questions

1 A car park is to be built in flexible materials. Private cars only will be admitted. Subgrade CBR is 5 per cent. What construction would you recommend? Explain your design and indicate why design charts based on a semi-analytical approach and intended for general use are of no help in such a case as this.

How would you modify your design if the car park also included access for delivery vehicles to a college canteen?

2 A temporary road is expected to carry 1000 standard axles during its life. Subgrade CBR is 4 per cent. Prepare two designs:

(**a**) one for general use by the public

(**b**) one for a road of restricted width which will be regularly used by large heavy vehicles following exactly the same path.

3 What factors should influence the choice of materials for the paving of a predestrianised shopping street? Discuss the relative merits and demerits of the more important materials available.

4 Describe how block paving derives its strength. How does a well-detailed design seek to provide a stable block paved surface?

Comment on the suitability of concrete block paving for

(**a**) a bus terminus

(**b**) a motorway

(**c**) a motorway service area car park.

5 Briefly describe the process by which concrete block paving is laid on a previously prepared base. Indicate which parts of the work are of particular importance in the production of a stable surface.

The designer has a choice of block thickness, bond and shape of blocks. For each of the types of work indicated below, indicate which block thickness, bond and block shape you consider to be appropriate. Give your reasons in each case.

(**a**) service road to a supermarket
(**b**) market square, access for vehicles up to 1 tonne gross weight only
(**c**) pedestrian area with service cables and mains beneath.

Further reading

BS 6717, Precast concrete paving blocks, Part 1, 1986: Paving Blocks, Part 2, 1989: Code of Practice for laying

BS 6677, Clay and calcium silicate pavers for flexible pavements, Part 1: specification for pavers, Part 2: code of practice for design of lightly trafficked pavements, BSI, 1986.

BS 7263, Precast concrete flags, kerbs, channels edgings and quadrants, Part 1: specification, Part 2: code of practice for laying, BSI, 1990.

Shakel, B., The performance of interlocking block pavements under accelerated trafficking, *Proc. 1st Int. Conf. on Concrete Block Paving*, 1980.

Barber, S. D. and Knapton, J., An experimental investigation of the behaviour of a concrete block pavement with a sand subbase, *Proc. Inst. Civ. Engrs*, Part 2, March 1980.

7. Highway drainage

7.1 Introduction

From earliest times arrangements have been made for the removal of surface water from paved areas. This is necessary because water lying on the surface is at best a nuisance for traffic and at worst a serious safety hazard, giving rise to a loss of adhesion between wheel and road and to such quantities of spray as will prevent adequate visibility.

The highway engineer is also concerned to ensure that adequate arrangements are made for the surface water runoff from a pavement since otherwise pools of standing water may form which will in time penetrate the structure of the pavement and cause premature failure. Damage may arise from weakening of the subgrade, from the washing away of fine material from unbound courses in the pavement, from the corrosive effects of moisture on elements of the pavement such as steel reinforcement, or in extreme weather from the action of frost on water in bound layers in the pavement.

For these reasons some arrangement must be made to ensure the adequate drainage of the surface of the road.

It may also be necessary to provide means for the removal of groundwater from the pavement foundation. This is discussed later.

The surface water drainage system for a highway should be efficient in respect of the collection of water from the surface, and in the conveying of surface water to a suitable discharge point. Before looking at these, a brief consideration of the elements of the drainage system is worthwhile.

7.2 Drainage components

The essential feature of a highway drainage system is that it should provide a route for water to flow along from the highway to a suitable discharge point, known as the outfall. This route may be at the surface (for example in the form of a ditch) or it may be underground in pipes or culverts (pipes of rectangular section and of a size which a man may comfortably enter). The outfall may be to another drain or sewer, a natural watercourse, a soakaway or some other convenient arrangement. Generally, pipes are used in preference to other means of conveying water from the road to the outfall.

It is clearly inconvenient to have a pipe opening in the surface of the road and so one of two arrangements is made. Either water is permitted to flow over the edge of the pavement and into a side ditch, or it is collected from the surface by means of gullies which connect either to a ditch or to a system of pipes leading to the outfall. The essential features of a ditch are that the side slopes should not exceed the natural angle of repose of the soil and that the depth should be such as not to allow the presence of surface water to raise the water table locally to an extent that would weaken the pavement. Gullies generally include a small sump into which detritus may sink for subsequent removal, preventing its entry into the drainage system where it may otherwise accumulate and cause blockages; and a 'trap', preventing the entry of floating material to the system. They may be positioned below the surface of the pavement, in which case water will enter through a grating placed flush with the surface (known as a channel inlet); or to one side of the pavement with an inlet mounted in the face of the kerb (known as a kerb inlet). Gratings and covers are in cast or ductile iron whilst the body of the gully is often preformed in concrete, clayware or plastic and is bedded in class 22.5/30 concrete. Typical construction details are shown in Fig. 7.1(a).

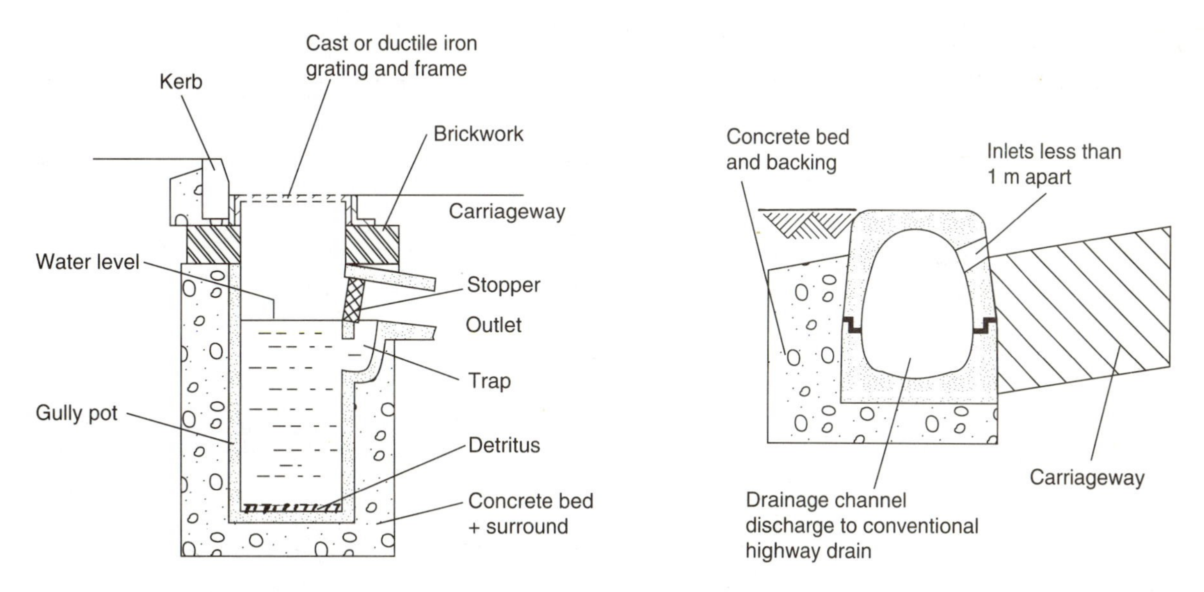

Figure 7.1 (a) Major elements of a typical gully. (b) A typical combined kerb and drainage block

Road gullies must be spaced at intervals along the road and this demands a longitudinal gradient so that water will flow to the gullies. Sometimes a gradient cannot be provided naturally and then an option which the engineer may choose is offered by combined drainage and kerb systems such as that shown in Fig. 7.1(b). These were introduced in the UK in the late 1970s and are often used to good effect.

Pipes will generally be of circular cross-section (rectangular and egg-shaped pipes are also available) and of clayware or concrete, although other materials may be considered on their merits. The structural design of the pipes will be dealt with later. Pipes are provided in units of between 900 mm and about 2000 mm long, depending on material and diameter, and in order that surface water should not leak into the subgrade it is important to ensure that joints between units are watertight.

A number of arrangements are available to achieve this, some of which are illustrated in Fig. 7.2. Whichever type of joint is chosen, the system should be subjected to a test upon completion to ensure that an adequate degree of watertightness has been achieved. Suitable tests are the air test and the water test. In the air test the pipe is stopped at all inlets and outlets, the empty system pressurised and the ability of the pipe run to maintain this pressure monitored with a manometer – a typical requirement being that a pressure capable of supporting a 75 mm column of water should be maintained for 5 minutes. In the water test the system is again stoppered at all entries but the uppermost, filled with water and a surcharge applied, typically in the form of a 1200 mm column of water, the system losing water through the joints at not more than a specified rate.

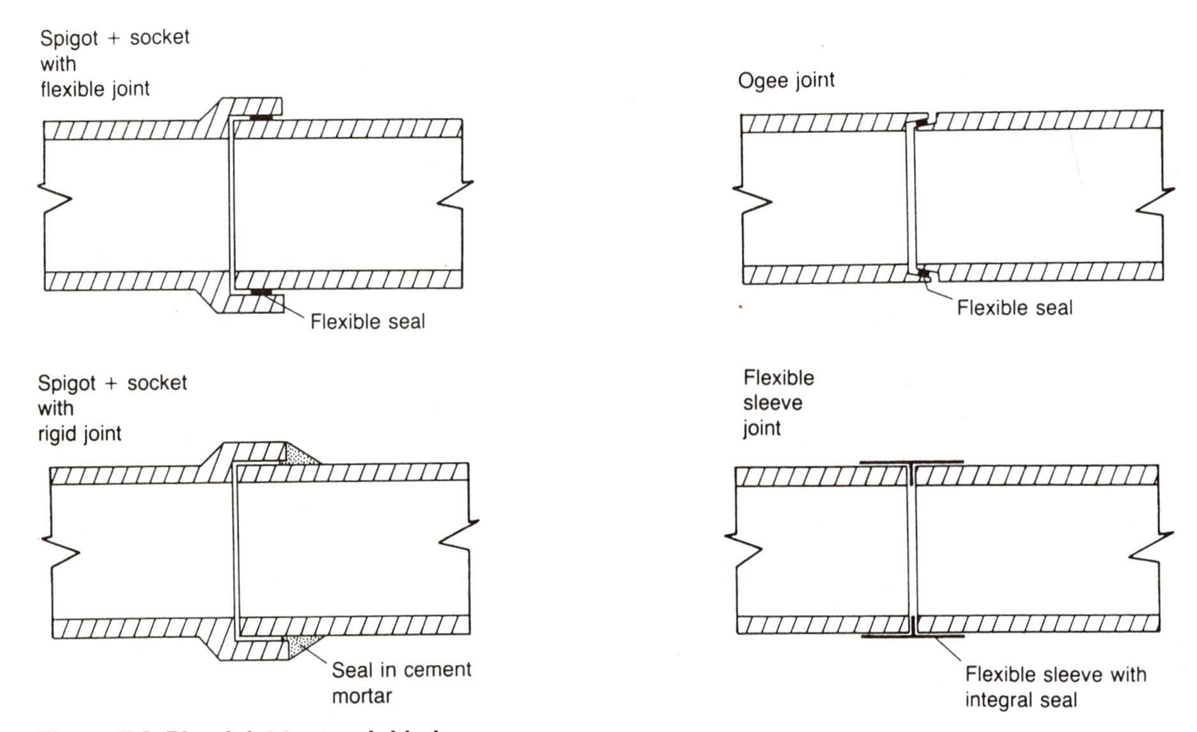

Figure 7.2 Pipe joint types: rigid pipes

Pipes will normally be laid in trenches prior to the construction of the road and, where the pipe lies beneath the pavement, it is particularly important that these trenches are refilled with material which can be satisfactorily compacted in a reasonable period, since settlement of this backfill material can have serious consequences. It is convenient to use material excavated from the trench as backfill where this is suitable, but cohesive soils in particular cannot be adequately compacted in the time usually available. In such cases imported granular material is often used, whether in the form of pea shingle or subbase or capping material. Exceptionally, where extra protection to the pipes is needed – as for example where they are laid very close to the formation – a concrete backfill may be specified. The structural effect of different pipe bedding types will be discussed later. Figure 7.3 illustrates a cross-section through a completed pipe.

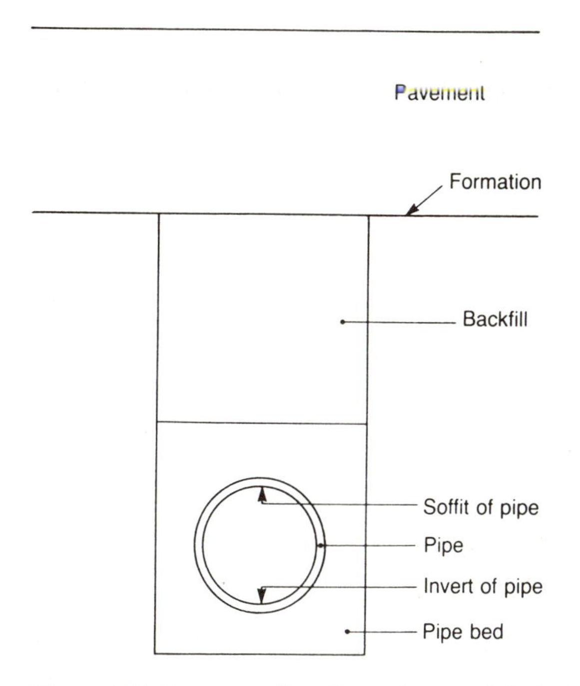

Figure 7.3 Cross-section through completed pipe laid in trench

To provide access for maintenance purposes, manholes are placed at intervals along the pipe. These are generally formed in precast concrete or engineering brickwork and must be substantually built to withstand earth pressures and to support highway loading. They must in addition provide adequate space for a man to work in safety and should, of course, be large enough to accommodate all pipes leading to the chamber. Manholes of the type shown in Fig. 7.4 will generally be adequate for highway drains whose depths do not exceed about 3 metres.

Catchpits are sometimes provided where water bearing a large proportion of suspended matter is expected to enter the system – for example where subdrains connect to the surface water system. Their function is to allow this suspended material to settle before it can be carried into the system where it may accumulate and cause blockages. Catchpits are similar in general construction detail to manholes, with the exception that a sump about 1000 mm

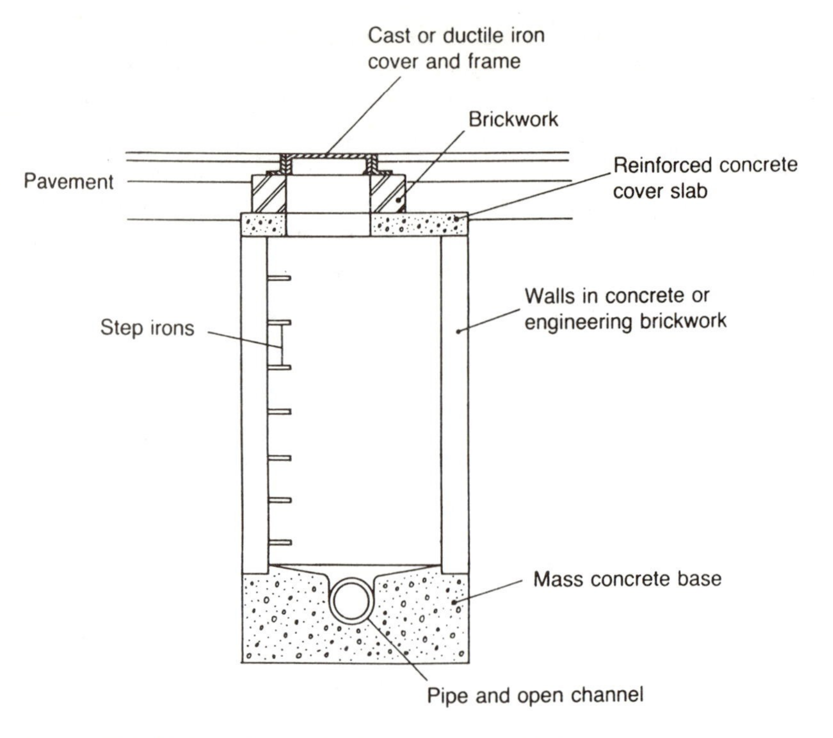

Figure 7.4 Major elements of a typical manhole

deep is formed below the invert level of the outgoing pipe. The depth of catchpits should not exceed about 3000 mm overall, or difficulty will be experienced in removing the settled material.

7.3 Design

The design process consists of devising an arrangement of pipes, manholes, gullies and so on which will be hydraulically and structurally sound.

7.3.1 Hydraulic design

The highway drainage system must be hydraulically effective in two respects. In order that the requirements of traffic and pavement engineering may be met, adequate arrangements must be made for the collection of water from the surface. Once water has entered the system it must be able to flow to the outfall; otherwise surface flooding may occur.

7.3.1.1 The collection of water from the surface

In all modern road construction, gradients are formed in the surface of the road to induce water to flow away from the trafficked areas. The most common arrangement is to provide a transverse gradient to quickly drain the wheel tracks. Where no kerbs are provided, water is then allowed to run over the edge of the road to an open channel of some sort whose design may be arbitrary or may consist of ensuring that sufficient outlets from the channel are provided. Where the road is kerbed a successful design will provide an appropriate combination of gully spacing and longitudinal gradient in the channel formed at the edge of the carriageway adjacent to the kerb. It is possible to form a channel remote from the kerb – for example in the middle of the paved area. This option is not attractive for traffic routes since the concentration of water in areas which vehicles may use at speed is unsafe, and the forming of a 'valley' cross-section may be dangerous for high-sided vehicles; but for car parks and other informal paved areas the arrangement may be worthwhile.

The gradients which may be formed in the surface of the road are subject to upper and lower bounds. The upper bounds are imposed not by hydraulic considerations but rather by the needs of traffic and are of the order of 1:15 transversely and perhaps 1:3 longitudinally. These gradients seldom occur in practice. Minimum gradients are more often a problem and arise out of the nature of the surfacing materials. It is the nature of the construction process that absolute uniformity cannot reliably be achieved; the regularity of various courses in the pavement is discussed in Chapter 4. In order that unavoidable irregularities in the surface do not result in standing water in trafficked areas it is necessary to limit the minimum gradients. Appropriate values are indicated in Table 7.1.

Table 7.1 Working minimum gradients: kerbed roads

	Minimum gradient	
Surfacing material	*Transverse*	*Longitudinal*
Rolled asphalt	1:50	1:200
Coated macadam:		
machine laid	1:50	1:120
hand laid	1:40	1:100
PQ Concrete	1:50	1:200
Block paving	1:35	1:100

Some difficulty can arise where the road is generally flatter than the minimum longitudinal gradient shown in Table 7.1. In such circumstances the designer may choose to create longitudinal gradients artificially in the channels of the road by raising and lowering these in relation to the general longitudinal gradient. Such an arrangement is known as false channel profiles.

In cases where the longitudinal gradient is small, an effective gradient may be achieved in the channel by varying the crossfall within the ranges indicated above. Outlets are provided at intervals along the road and the channel is shaped to drain to these from intermediate high spots. If the road is horizontal, or very nearly so, and if there is a minimum acceptable gradient, it follows that the spacing of the outlets will be a function of the maximum acceptable difference in level between the high and low spots. This difference in level cannot be varied at will since there is a maximum and minimum acceptable kerb face – commonly 150 mm and 75 mm respectively – and since where there is a kerb this is laid in such circumstances parallel to the centre line of the road rather than undulating between the high and low spots in the channel. An undulating kerb rarely presents an acceptable appearance.

Figure 7.5 shows a longitudinal section through such a channel.

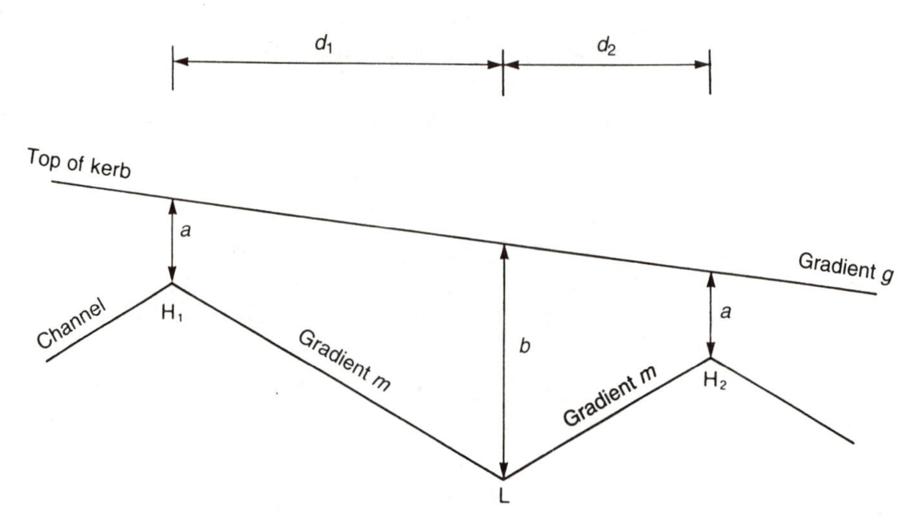

Figure 7.5 False channel profile

If a = minimum kerb face
b = maximum kerb face
m = minimum acceptable gradient expressed as a fraction
g = gradient of centreline expressed as a fraction
RL(L) = reduced level at L, etc,

then $RL(L) = RL(H1) + a - g \cdot d_1 - b = RL(H) - m \cdot d_1$ where d_1 and d_2 are as shown on Fig. 7.5.

Therefore $b - a = (m - g) \cdot d_1$

and
$$d_1 = \frac{b-a}{m-g}$$

Similarly
$$d_2 = \frac{b-a}{m+g}$$

from which
$$\frac{d_1}{d_2} = \frac{m+g}{m-g}$$

and
$$\frac{d_1}{d_2}$$

approaches infinity as g approaches m. Using this simple geometry the spacing of high and low spots can easily be calculated. Note that if

$a = 75$ mm

$b = 150$ mm

$m = 1{:}200$

and $g = 0$

then the spacing between peaks becomes 30 metres. This gives an unacceptable ride at high speed, in which case alternative solutions such as over-edge drainage may be sought.

Another area in which difficulty can arise is that between a road of symmetrical cambered cross-section and a superelevated curve (see 8.2.4.1), where the transverse gradient on the half carriageway width on the outside of the curve changes from draining down from the centre line to the outer kerb to draining down from the outer kerb, to the centre line and then to the inner kerb. There is, of necessity, an area here where there is no transverse gradient, and the designer should ensure that superelevation is not introduced so gradually as to create large almost flat areas of carriageway, nor so sharply as to cause discomfort to road users or to kink the edge of the carriageway. Common practice in such cases is to ensure that the carriageway edge profile does not vary in gradient by more than about 1 per cent from the general alignment of the road, and by ample smoothing of all changes in edge profile.

Once an adequate gradient has been provided, both transversely and along the channel, we must determine the gully positions.

Clearly, a gully should be provided at every low spot along the channel. The positions of these should be determined from the geometric design. A gully should also be provided in the outer channel of a cambered road immediately before it changes to superelevation or to a straight crossfall from one side to the other – in order that water flowing along the channel shall be prevented from flowing across the road. Elsewhere, the spacing of the individual gullies is determined from a consideration of the capacities of the individual gullies and of the width of flowing water in the channel.

The efficiency of a gully – measured as the proportion of water flowing towards the gully along the channel which actually enters the gully – will depend on the entry conditions at the gully. Narrow flows and low flow velocities lead to good gully performance as the tendency of water to overshoot the gully is reduced. Channel inlet types are generally much more efficient than kerb inlet gullies, provided that both are adequately maintained. Commercially available heavy duty channel entry gullies can achieve efficiencies in excess of 95 per cent in all but the most demanding circumstances whereas in the same conditions kerb inlet types may achieve only 30 per cent or less.

The width of flow in the channel will depend on the cross-sectional shape of the channel, its gradient, the intensity of the rainfall, and the spacing of the gullies.

Manning's equation[1] has been modified by Russam[2] to the approximate form:

$$Q(\text{approx}) = 0.315 \cdot \frac{K}{n} \cdot F^{8/3} \cdot T^{5.3} \cdot L^{1/2} \quad [7.1]$$

in the special case of the flow at the edge of a kerbed road, in which equation

Q is the flow in unit time

K is a constant equal to unity when the unit of length is the metre

n is Manning's roughness coefficient, often equal to about 0.01 in this application

F is the width of flow

T is the transverse gradient, expressed as a fraction

L is the longitudinal gradient, expressed as a fraction.

This relationship may be used to enable the width of flow to be predicted if the rate at which water drains to the channel is known. This quantity, the runoff, may be calculated from a knowledge of the size of the area drained and the intensity of rain falling on that area. Rainfall intensity is predicted on a statistical basis and is discussed below; for the sake of simplicity here it is assumed that the design rainfall intensity will be 50 mm/hr. Flow in the channel will clearly be maximised immediately upstream of the gully.

By controlling the gully spacing the designer may therefore control the width of flow of water in the channel and contain this within whatever limits are considered to be acceptable. Typically one may design for maximum flow widths of 0.5 metres in roads next to main pedestrian routes and in high speed roads where spray may be a particular nuisance, and perhaps 1.0 metres in minor roads and car parks, where pedestrian and vehicle activity is likely to be small in the severe weather conditions represented by a storm of 50 mm/hr intensity.

The charts in Fig. 7.6 indicate maximum gully spacing for various combinations of transverse and longitudinal gradients consistent with a storm intensity of 50 mm/hr, channel entry gullies and a width of flow of 500 mm. For other intensities the spacing may be varied proportionately and where kerb inlet gullies are to be used the spacing should be reduced by a factor of three. Gullies should be placed at the indicated maximum spacing along the road, working away from low spots and other constraints.

The approach will not work in cases where the gradients indicated in Table 7.1 are not achieved. Level or nearly level roads can only be adequately drained by the provision of overedge drainage, discharging usually to an open channel, or by accepting a considerably increased width of flow in a kerbed channel. Ditches cut to a 'V' shape have already been discussed and are usually designed in an arbitrary manner. For major roads this approach may not be sufficient; particularly in the case of motorways where the drainage of large impermeable areas may be required.

Two alternatives present themselves in the drainage of a level road of motorway character; overedge drainage to a separate drainage channel, or drainage to a kerbed hard shoulder. In either case the design problem consists of balancing the capacity of the channel (either separate or in the hard shoulder)

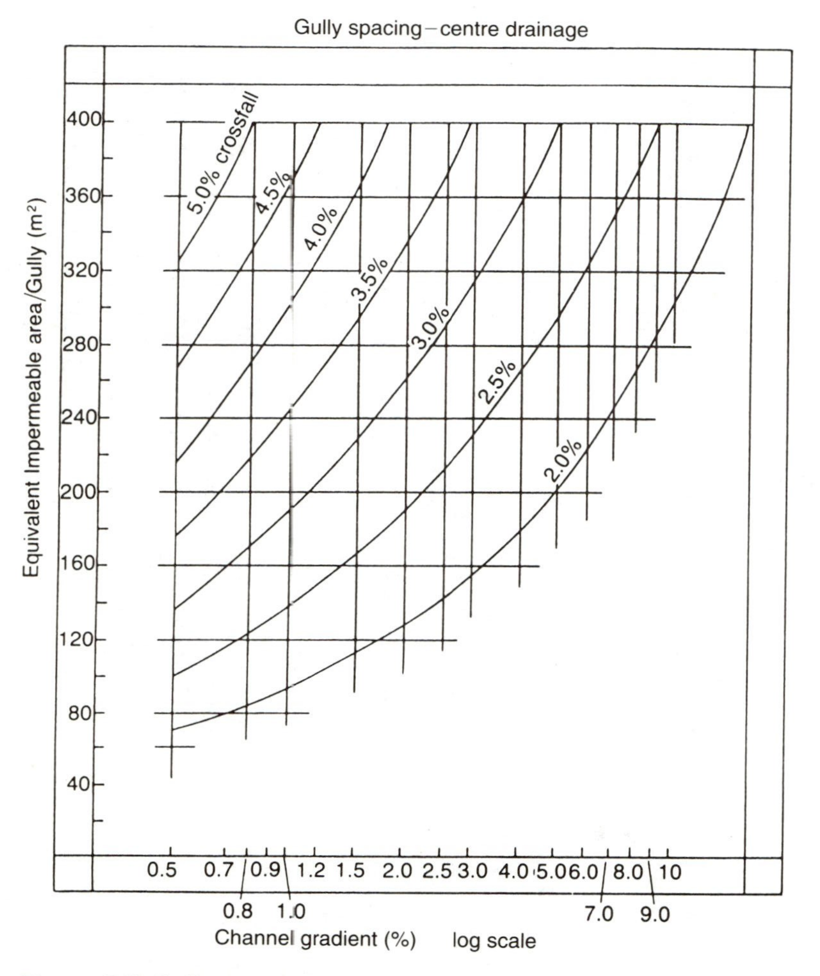

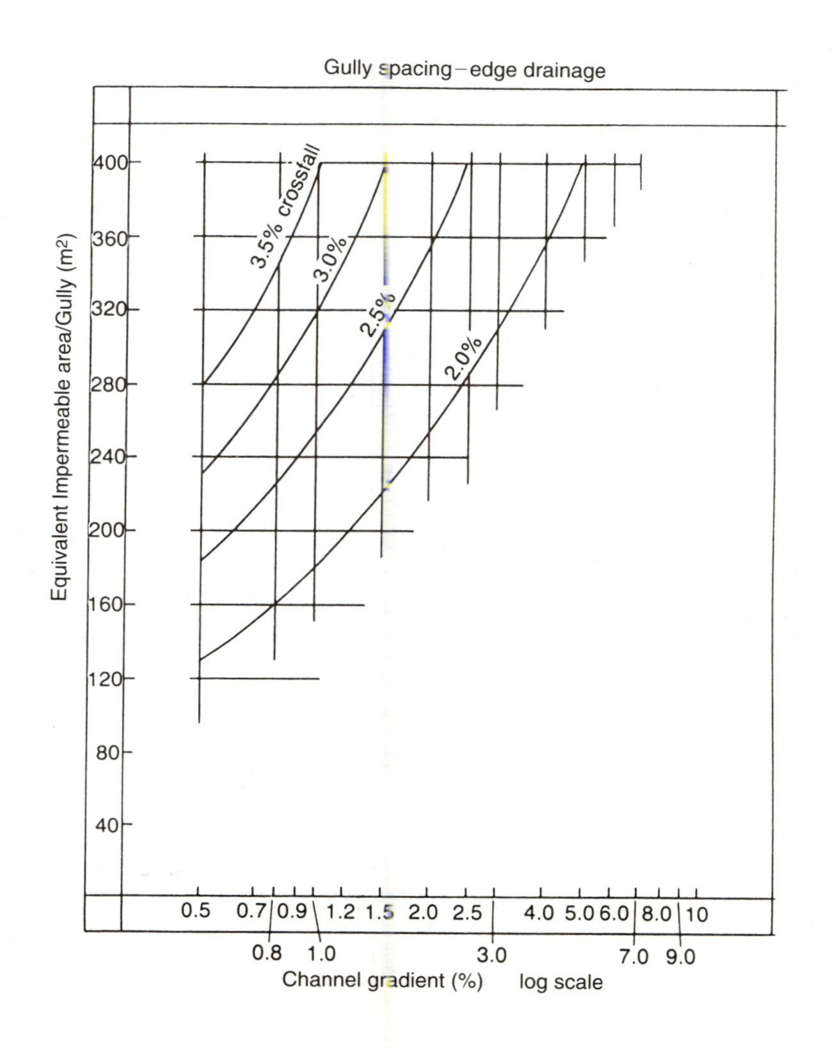

Figure 7.6 Gully spacing

and the spacing of the outlets from that channel so that the amount of water in the channel is not excessive. If the maximum outlet spacing in metres is represented by J then in the case of a trapezoidal channel of constant cross-section

$$J = \frac{0.235(S + 0.5 \cdot G \cdot H)^{12/13} \cdot H^{16/13}}{(IW)^{10/13}}$$

in which

S is the width of the bottom of the channel (mm)

G is the factor by which the width of the trapezoidal channel increases with increased distance from the bottom

H is the depth of the channel (mm)

I is the rainfall intensity (mm/hr)

W is the width of the paved area (m)

and in the case of drainage along the hard shoulder

$$J = 545 \cdot \frac{(F^3)^{3/4}}{(IW)} \cdot (100T)^{23/16} \cdot \left(1 + \frac{BF^{7/4} \cdot (100L)^r}{(IW)^{7/8}}\right)$$

in which

F is the maximum flow width

r is an index such that $r = 2.32 - 13L$

B is a coefficient related to T such that

$T = 0.01\ 0.015\ 0.02\ 0.025\ 0.03\ 0.04\ 0.05$

$B = 190\ 265\ 326\ 380\ 416\ 448\ 448$

T and L are the transverse and longitudinal gradients,

and the other variables are as before.

7.3.1.2 Runoff

In either case, the required capacity of the outlet can be calculated from a consideration of the area draining to that outlet and the intensity of rainfall.

The peak runoff which may be expected from a catchment area may be calculated[3] from the equation

$$Q = 3.61CAI$$

in which

C is a volumetric runoff coefficient whose function is to reflect the loss of some rainfall through cracks and into depressions and by drainage onto pervious areas. Values of C range from about 0.6 on catchments with rapidly draining soils to about 0.9 on catchments with heavy soils; for the drainage of wholly paved highways take $C = 0.95$.

A is the impermeable area within the contributing catchment (ha)

I is the average rainfall intensity during the time of concentration (mm/hr)

Q is the peak runoff (litres/sec).

If A is expressed in square metres then this equation becomes

$$Q = 3.43AI \times 10^{-4} \quad [7.2]$$

7.3.1.3 Conveying water from the site

The essential problem in the hydraulic design of the system of pipes taking water to the outfall is to decide the intensity of rainfall for which the system is to provide.

Records of rainfall have been kept throughout the UK for many years and it has been found that high rainfall intensities occur more frequently in storms of short durating than in long ones, and that storms of high intensity occur at less frequent intervals than do storms of low intensity. If records for a typical two-year period were plotted on a chart such as Fig. 7.7 it would be possible to drawn a curve of the hyperbolic form shown which would just include all the storms during that period.

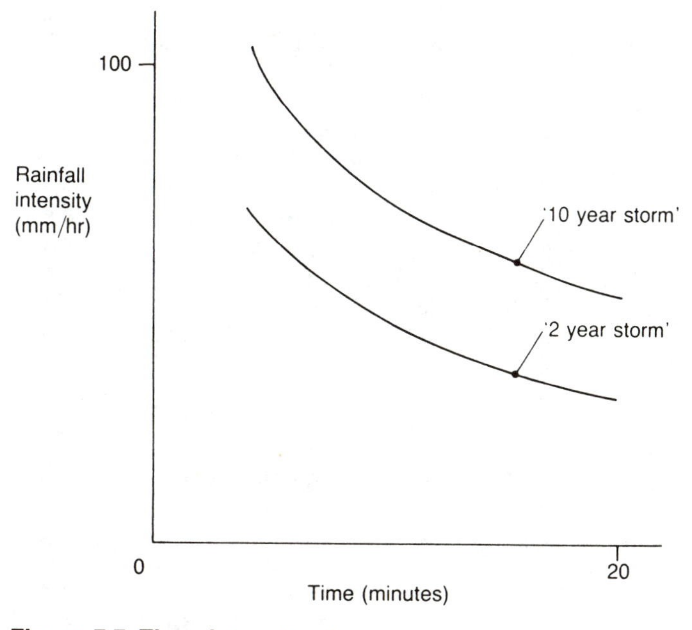

Figure 7.7 Time–intensity diagram

If a typical 10-year period were then to be considered, some storms would probably be found to have occurred at an intensity greater than that found in the typical two-year period. In plotting these as before it will be necessary to draw a second hyperbola to include all the storms which took place during the longer period, as shown in Fig. 7.7. It is possible to produce a family of curves representing a range of such return periods. Various attempts have been made to model this behaviour mathematically. One of the most widely used equations in the UK is that known as the Bilham formula, subsequently updated by the Meteorological Office,[4] in which

$$I = \frac{60}{t}[(202.26Nt)^{0.2817} - 2.54] \qquad [7.3]$$

where

N is the storm return period expressed as once in N years
t is the time in minutes since the start of the storm
I is the rainfall intensity at time t, measured in mm per hour.

Values of I for ranges of N and t likely to be appropriate for most highways applications are given in Table 7.2.

Table 7.2 Rates of rainfall in millimetres per hour for the UK

Time (min)	Return period (years) 1	2	5	Time (min)	Return period (years) 1	2	5
4.0	55.6	69.3	87.9	7.0	43.3	54.6	70.6
4.1	55.1	68.6	87.1	7.2	42.7	54.1	69.9
4.2	54.6	68.1	86.4	7.4	42.2	53.3	68.8
4.3	54.1	67.3	85.6	7.6	41.7	52.6	68.1
4.4	53.6	66.8	85.1	7.8	41.1	51.8	67.3
4.5	53.1	66.3	84.3	8.0	40.6	51.3	66.6
4.6	52.6	65.5	83.6	8.2	39.9	50.6	65.8
4.7	52.1	65.0	83.1	8.4	39.4	50.0	65.0
4.8	51.6	64.5	82.3	8.6	39.1	49.5	64.3
4.9	51.3	64.0	81.5	8.8	38.6	48.8	63.8
5.0	50.8	63.5	81.0	9.0	38.1	48.3	63.0
5.1	50.3	63.0	80.5	9.2	37.6	47.8	62.2
5.2	49.8	62.5	79.8	9.4	37.1	47.2	61.7
5.3	49.5	62.0	79.3	9.6	36.6	46.7	61.0
5.4	49.0	61.5	78.7	9.8	36.3	46.2	60.5
5.5	48.8	61.0	78.0	10.0	35.8	45.7	60.0
5.6	48.3	60.5	77.5	10.5	34.8	44.5	58.4
5.7	48.0	60.0	77.0	11.0	34.0	43.4	57.2
5.8	47.5	59.7	76.5	11.5	33.0	42.4	55.9
5.9	47.2	59.2	76.0	12.0	32.3	41.4	54.6
6.0	46.7	58.7	75.4	12.5	31.5	40.4	53.3
6.2	46.0	57.9	74.4	13.0	30.7	39.4	52.3
6.4	45.5	57.2	73.4	13.5	30.0	38.6	51.3
6.6	44.7	56.1	72.4	14.0	29.2	37.9	50.5
6.8	43.9	55.4	71.6	14.5	28.7	37.1	49.3

An alternative to equation [7.3] exists in the modified rational method on which the suite of computer programmes known as the Wallingford Procedure is based.[3] This approach is derived from a study of storm characteristics in the area concerned and gives superior results to equation [7.3] for large catchments; for all but the major highway schemes the above equation will give adequate results.

In areas other than the UK it may be expected that equation [7.3] will not necessarily apply. If suitably detailed local records are not available, the equations developed by Bell[5] may be of value.

Bell found that if the rainfall depth over t minutes unlikely to be exceeded during N years if P^t_N then

$$P^t_N = (0.21 \text{ Ln } N + 0.52)(0.54t^{0.25} - 0.50)P^{60}_{10}$$

where Ln is the natural logarithm.
Bell also provided means of estimating P^{60}_{10}:

$P^{60}_{10} = 0.27MN^{0.33}$ ($M = 50$ or less)

$P^{60}_{10} = 0.97M^{0.97}N^{0.33}$ (M between 50 and 115)

where M is the mean of the maximum annual observational-day precipitation (mm)
N is the mean annual number of rainfall days (N between 1 and 80).

7.3.1.4 Time of concentration

In the design of surface drainage for highway purposes it is assumed that the maximum runoff from an area occurs when the runoff is calculated on the basis of a storm duration equal to the time of concentration, t_c, of the area. The time of concentration is the time which must elapse from the start of the storm until water has flowed from the whole of the area upstream of the point in question, to that point, and is generally taken to be such that

$$t_c = t_e + t_f \qquad [7.4]$$

in which

t_e is the time of entry to the system – that is, the time taken for water falling on the most remote part of the catchment to flow across the surface, into a gully or equivalent, and through the gully connecting pipe to the main highway drain

t_f is the time of flow along the main highway drain to the point in question.

At times earlier than this, not all the catchment is contributing to the runoff; at times later than this, the rainfall intensity is reduced. Values of the time of entry have been recommended[3] as shown in Table 7.3. Time of entry depends on the nature of the catchment; small steep catchments are fast-draining and therefore have low values of t_e whereas large, flat areas drain relatively slowly.

Table 7.3 Recommended time of entry to highway drainage systems

Storm return period (years)	*Time of entry (t_e minutes)*
5	3–6
2	4–7
1	4–8

Figure 7.8 is derived from equations [7.1] and [7.3] and relates distance to time in the case of flow along road-edge channels under various conditions and may be used as a rough guide to the assessment of t_e. To the values indicated in the figure should be added an allowance for the flow over the surface to the channel and along the pipe collecting the gully or equivalent to the main highway drain.

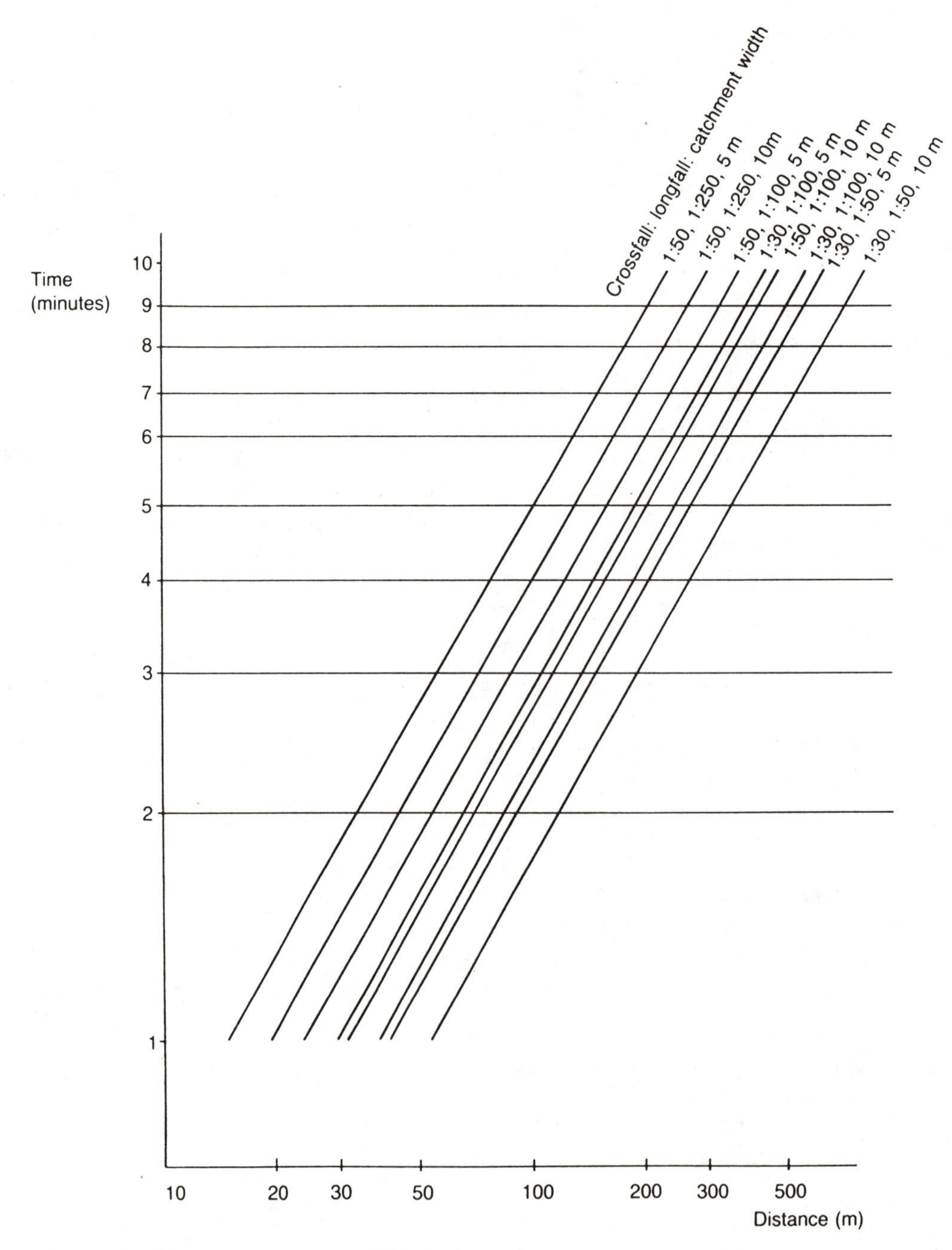

Figure 7.8 Time of flow in road-edge channels

7.3.1.5 Selection of storm return period

A return period storm is one which is unlikely to be exceeded in intensity during a stated time, which time is known as the return period. If the return period under consideration is increased, the intensity of the storm will also increase. In dealing with increasing return periods, the designer finds himself dealing with events which are progressively less likely to happen, and which demand progressively more expensive measures to be taken to provide a drainage system capable of dealing with the consequences of those events. At any particular return period the designer may wish to decide whether the extra cost of providing a further increased level to service (by designing for a greater return period) can be justified in terms of the reduced inconvenience experienced by road users. Figure 7.9 represents the relationship between drainage costs, user disbenefits and storm return period. The optimum return period is that at which the disbenefits are equal to the drainage costs, and clearly depends very much upon the form of the disbenefit curve.

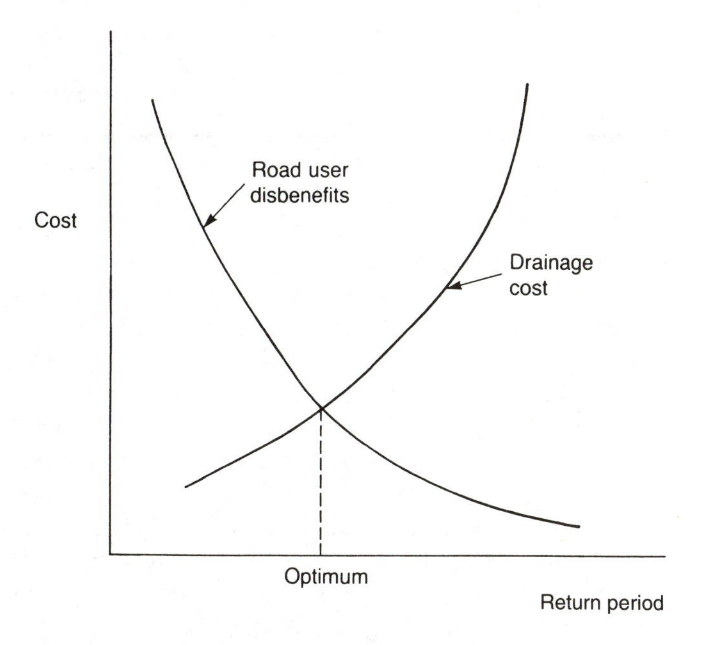

Figure 7.9 Return periods and costs

In the case of main surface water drainage, the consequences of flooding may be particularly severe and a large return period of many years may be appropriate. Infrequent flooding of drains which serve the highway alone does not necessarily lead to damage of the same order of severity and a lower return period can be used. For example, the Thames Barrier in London, intended to protect the centre of the capital from flooding, was designed for a 1000 year period; whereas for general use in the UK the return periods shown in Table 7.4 are recommended in the absence of more positive guidance.

Table 7.4 Recommended storm return periods for use in the UK

Nature of site	*Return period*
Sites with average ground slopes greater than 1:100	1 year
Sites with average ground slopes 1:100 or flatter	2 years
Sites where the consequences of flooding will be severe	5 years

7.3.1.6 Design of individual pipe lengths

It is the practice to design each length of pipe between one manhole and another as a separate part of the whole drainage system. In order that this may be achieved, an awareness of the requirements for positioning manholes is necessary. Pipe lengths are generally laid straight between manholes, and are usually arranged to drain by gravity.

The primary function of manholes is to provide access for maintenance and they are positioned accordingly. Access is required at all changes in direction throughout the length of the drain, at the head of each branch, at junctions between two or more drains, where the size of the pipe is changed, and on long straight uniform sections at intervals of about 100 metres. These standards may be relaxed somewhat on culverts – but pipes of this size are unlikely to concern the designer of highway drains.

Some thought should also be given to the location of manholes in relation to other highway features. In Chapter 5 it was shown that where a jointed concrete pavement is used the bay layout should complement the positioning of manholes. It will of course be necessary in planning the route of a drain to ensure that there is no conflict between the drain and other underground features such as service pipes and cables or building foundations; trial excavations are often worthwhile in areas of uncertainty. Once the road has been opened to traffic, regular maintenance of the drainage system will be encouraged if access to the manholes is convenient and for this reason it is prudent to avoid placing manholes in positions such that major traffic diversions will be necessary before they can be entered.

To prepare the design, the engineer will first indicate on a plan of the road the manhole positions that are required. Pipe lengths between the manholes are numbered according to the convention shown in Fig. 7.10; pipes in the longest run from head of pipe to outfall being numbered 1.01, 1.02 and so on, and the branches from this pipe run numbered sequentially from the top of the main run 2.01, 2.02 etc. in the case of branch 2, 3.01, 3.02 in branch 3 and so on. Each individual pipe length between manholes is then designed as below.

The design routine may conveniently be broken down into a number of stages.

1 *Measurement of the impermeable area*

This measurement is made from plans of the finished highway. Impermeable

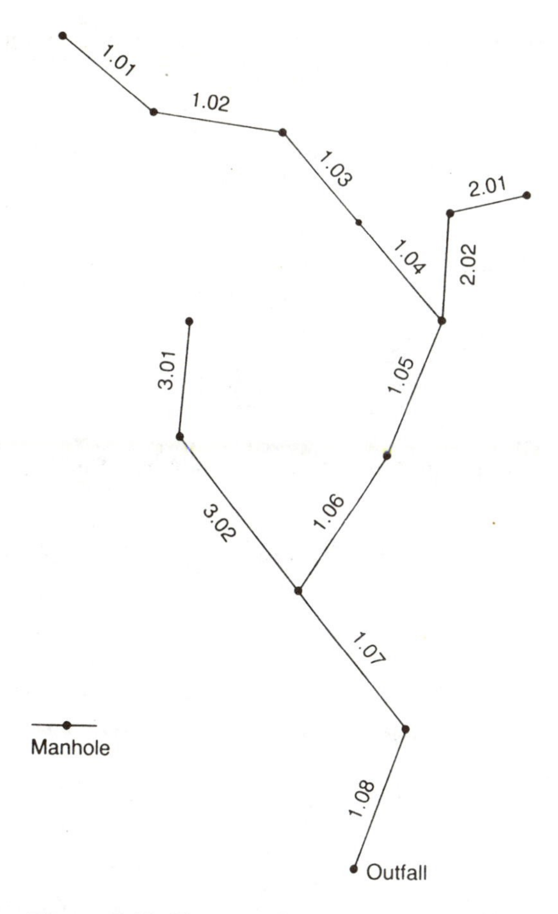

Figure 7.10 Pipe numbering convention

areas are either measured individually throughout the catchment, or, in the case of large uniform catchments, the proportion of the whole area which is effectively impermeable may be calculated for a representative sample and applied globally. The total area draining to each pipe should be determined.

2 *Determination of rainfall intensity*
This has been discussed. Equation [7.3] may be used for works in the UK, the time of concentration being established initially from an estimate of the pipe size and gradient (and therefore time of flow) that will be required. Local data or Bell's equations may be used elsewhere.

3 *Selection of pipe size*
An initial assumption is made that the pipe will be the greater of the smallest acceptable (often 150 mm diameter), and the same size as that immediately upstream. The time of flow in the pipe may be based upon the full-bore velocity of flow in the pipe, an approximation which is often considered to yield adequate results. Charts and tables are available relating velocity of flow to

pipe size, gradient and pipe material which are based on the Colebrook–White equation

$$V = -2\sqrt{(2gDi)}\log\left(\frac{k}{3.7D} + \frac{2.51f}{D\sqrt{(2gDi)}}\right)$$

in which

V is the velocity of flow in the pipe

g is the gravitational acceleration

i is the hydraulic gradient

f is the kinematic viscosity of the fluid

D is the pipe diameter

k is a linear measure of effective roughness, for example $k=0.6$ mm for mature surface water sewers

Figure 7.11 shows a chart indicating results obtained from this equation (see reference 6).

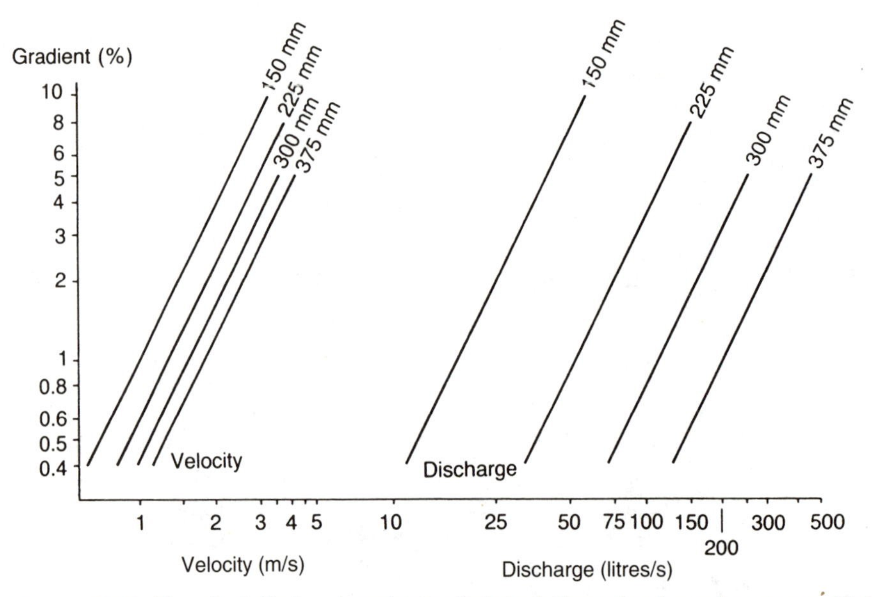

Figure 7.11 Flow in full circular pipes of stated diameter (roughness coefficient 0.6 mm)

It is necessary to ensure that the pipe has adequate capacity for the expected runoff, and that the velocity of flow in the pipe is not too low. Surface runoff from highways may be expected to contain a large amount of suspended solids which will settle out if the water flows too slowly and may then cause silting up of pipes and manholes. To avoid this, velocities less than 0.75 metres per second are usually avoided. Where the entry of silt to the system is expected to be particularly heavy, a catchpit may be provided.

In calculating gradients, allowance should be made for the effective loss of level at manholes in which pipe sizes increase; this change to invert level arises from the usual practice of making soffit to soffit connections. Invert to invert connections are not favoured since they often prevent the full capacity of the larger pipe being made available; the water cannot rise up to meet the soffit level downstream of the manhole.

Other factors which affect gradient include the level of the outfall and the topography of the site. It is generally convenient to follow surface levels as much as possible and to arrange matters so that the depth of cover above the pipe is about 1.5 metres – this avoids the structure of the pavement while not causing excessive cost. Construction costs increase with increasing depth and it is not attractive to provide pipes deeper than about 2.5 metres to invert if this can be avoided. If pipes are laid at insufficient depth they may be damaged by the construction of the road and may be found to be too shallow to be effective since standard gully outlets will be too low to connect to the main pipe. Where pipes are unavoidably laid near to the formation and where this arrangement is hydraulically acceptable it will usually be necessary to provide special protection to the pipes, perhaps in the form of a concrete bed and surround. In some cases it is necessary to lay pipes above the formation – perhaps where the road is to be built over areas of fill – in which case all earthworks should be completed before construction of the highway drainage system is begun.

The design process consists in outline of:

1 estimating the pipe size
2 from a knowledge of the gradient and length of the pipe, predicting the time of flow
3 adding this time of flow to the time of entry for the system, or in the case of a pipe not at the head of the run, adding this time of flow to the time of concentration of the previous pipe. The time of concentration at the point in question is thus determined
4 predicting the rainfall intensity from equation [7.3] or by other means
5 calculating the runoff from the whole catchment draining to the point in question, or into those pipes draining to it, from equation [7.2]
6 comparing this runoff with the capacity of the pipe at the gradient proposed, and adjusting the pipe size or proposed gradient as necessary.

A calculation sheet which clarifies the design process is shown in Fig. 7.12.

7.3.2 Structural design

Having designed the pipe locations and diameters, it remains to ensure that the pipes are of adequate strength and sufficiently well protected that they do not fail in use. Elements of this aspect of the design process are the selection of a pipe type and the choice of material for pipe bedding. Trench backfill must also be considered.

Problem: design the highway drain for a road 340m long with 1×7·3m carriageway and 2×2m wide footways, no verges. Longitudinal gradient = 1%. Carriageway is cambered, gradient down from centre line is 3% on each side. 2 year storm return period. Pipe k_s=0·6. Hydraulic design only required.

① Gully spacing

From Fig. 8·6, provide one gully per 320m² impermeable area

$$\text{Spacing along road (each side)} = \frac{320}{\frac{7{\cdot}3+2}{2}} = 57\ \text{m}$$

Provide gullies, pipes and manholes thus :– (NTS)

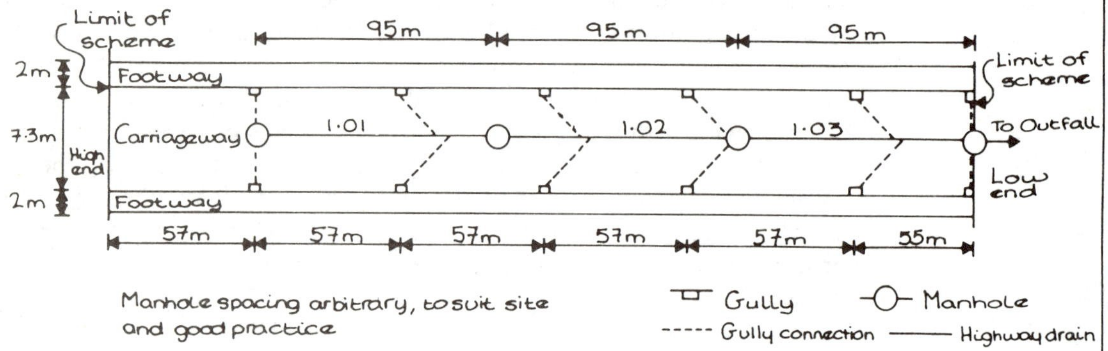

Limit of scheme
95m | 95m | 95m
2m Footway
7·3m Carriageway
High end
2m Footway
1·01 | 1·02 | 1·03
To Outfall
Low end
Limit of scheme
57m | 57m | 57m | 57m | 57m | 55m

Manhole spacing arbitrary, to suit site and good practice

Gully — Manhole
----- Gully connection —— Highway drain

② Time of entry, t_e, at top end = time of flow on road to channel + time of flow in channel (fig. 8·8 gives about 2·2 min) + time of flow in gully + connection = say 3 minutes in this case.

③ Sizes of catchments - if the areas shown in the sketch are to be drained, and no others, these are :–

PIPE	CATCHMENT AREA
1·01	(57+57)×(2+7·3+2) = 1288 m²
1·02	1288 + 57×11·3 = 1932 m²
1·03	1932 + (2×57×11·3) = 3220 m²

④ Pipe capacity design : Assume pipes are laid parallel to the surface - i.e. at 1% gradient

Pipe Nº	Length (m)	Gradient (%)	Trial size (mm)	Flow velocity (m/s)	Capacity (l/s)	t_e (mins)	t_f (mins)	t_c (mins)	Rainfall intensity (mm/hr)	Catchment area (m²)	Runoff (l/s)	OK?
1·01	95	1	150	1·0	18·8	3	1·6	4·6	65·5	1288	28·9	No - pipe too small
1·01	95	1	225	1·3	50	3	1·2	4·2	68·1	1288	30·1	Yes : Capacity > Runoff
1·02	95	1	225	1·3	50	4·2	1·2	5·4	61·5	1932	40·8	Yes
1·03	95	1	225	1·3	50	5·4	1·2	6·6	56·1	3220	62·0	No
1·03	95	1	300	1·55	105	5·4	1·0	6·4	57·2	3220	63·2	Yes
References →				Fig. 8·11		Pipe 1·01 : ②			Table 8·2	③	Eqn 8·2	

Conclusion : Use 225mm for 1·01, 1·02 ; 300 mm for 1·03

Figure 7.12 Simplified drainage calculation

7.3.2.1 Pipe types

The service conditions of a highway drain may include external loading due to earth pressures and surcharges imposed by the road itself and its traffic, scour and wear due to the passage of suspended particles in the runoff water, and chemical attack from within by de-icing salts and spillages in the road and from without by aggressive chemicals such as acids and some sulphates present in the soil. The pipe should therefore either be of a material which can withstand these conditions, or be protected from them.

Pipes are commonly made of clayware, cast iron, steel, pitch fibre or plastic. Of these, cast iron and steel are prohibitively expensive and are seldom used for pipe lengths which drain by gravity.

The material on which the pipe rests is known as its bed. This material may be placed under the pipe only, or may be extended up to half the pipe depth (bed and haunch), or to completely cover the pipe (bed and surround). Depending on the nature of the material and the bedding arrangement chosen, the pipe will to a greater or lesser extent be protected from external loads. The designer may therefore choose between the relative benefits of providing a strong bedding and weak pipe, or vice versa.

Bedding materials in common use are concrete (plain or reinforced), pea shingle (single sized granular aggregate of 14 mm or 20 mm nominal size), sand, or the material previously excavated from the trench. So that subsequent settlement of unbound bedding and backfill materials may be avoided, it is important that these should be fully compacted. For unbound materials, ease of compaction is a matter of particular importance since compaction in a trench working around a pipe is often awkward. Ease of compaction may be measured by the Compaction Factor test in which a standard cylinder, 250 mm long and 150 mm in diameter, is filled with the uncompacted material, this material then being removed and returned to the cylinder in fully compacted quarters. The compaction factor is defined as

$$\text{Compaction factor} = \frac{\text{Height of tube} - \text{Depth of compacted material}}{\text{Height of tube}}$$

and is such that materials with a compaction factor of 0.1 or less are generally suitable for use as a pipe bedding material; those whose compaction factor is between 0.1 and 0.3 are suitable only in dry conditions, while those whose compaction factor is greater than 0.3 are always unsuitable.

Rigid pipes are better able to withstand external pressures than are those made of flexible materials, and so where loads are large the bedding types for the former offer less protection than do those for the latter. Soil pressures increase with increasing depth, whereas pressures due to traffic or other surcharge at the surface rapidly reduce with increasing depth. Some standard bedding types for pipes made of rigid materials (clayware and concrete) are shown in Fig. 7.13(a) and those for pipes made of flexible materials (pitch fibre, plastics) are shown in Fig. 7.13(b). Typical applications for each are shown in Fig. 7.13(c).

It is possible to provide rigid joints between pipes made of rigid materials but to do so can lead to the pipes being overstressed as a result of ground movements after construction. Most rigid pipes are therefore provided with flexible joints (see Fig. 7.2) in order that a small amount of relative movement can be accommodated between one unit and the next, and in cases where a rigid (concrete) pipe bedding is used it is important to ensure that the required flexibility is maintained. This is achieved by providing movement joints in the concrete bed at intervals of about 5 metres, placed to coincide with pipe joints. The movement joints consist simply of a collar in fibreboard or similar compressible material, fitted around the pipe prior to concreting and arranged to form a complete discontinuity in the concrete.

7.4 The effects of subsurface water

By the term 'subsurface water' is meant water contained in the soil. Its source is generally surface water but it may arise elsewhere – for example, from

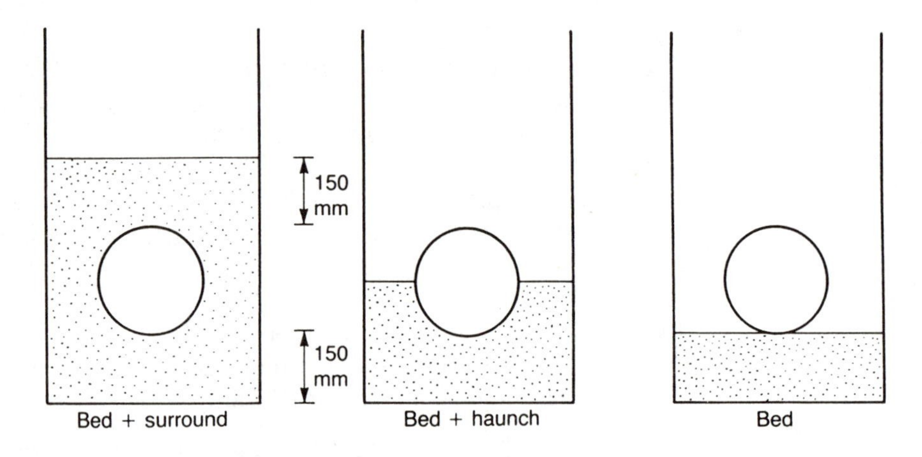

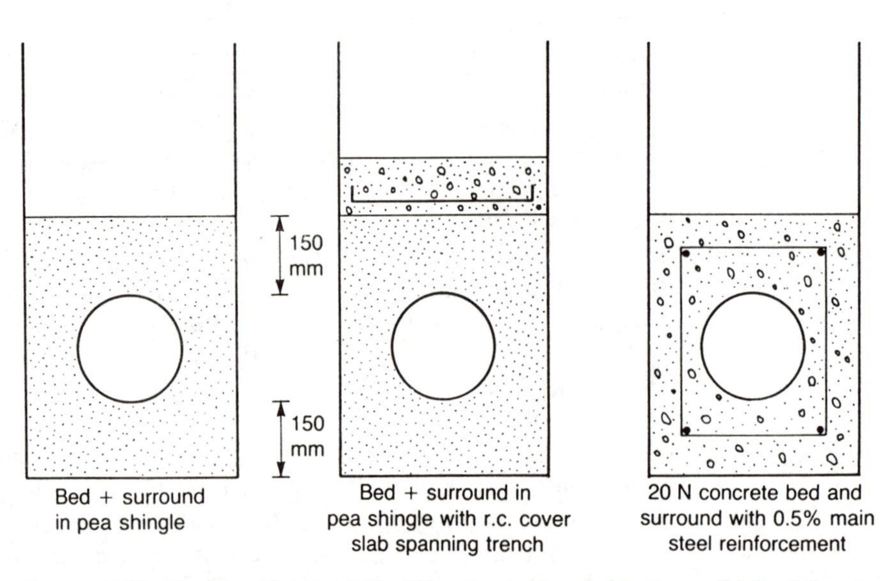

Figure 7.13 Pipe bedding. (a) Bedding types for rigid pipes. (b) Bedding types for flexible pipes.

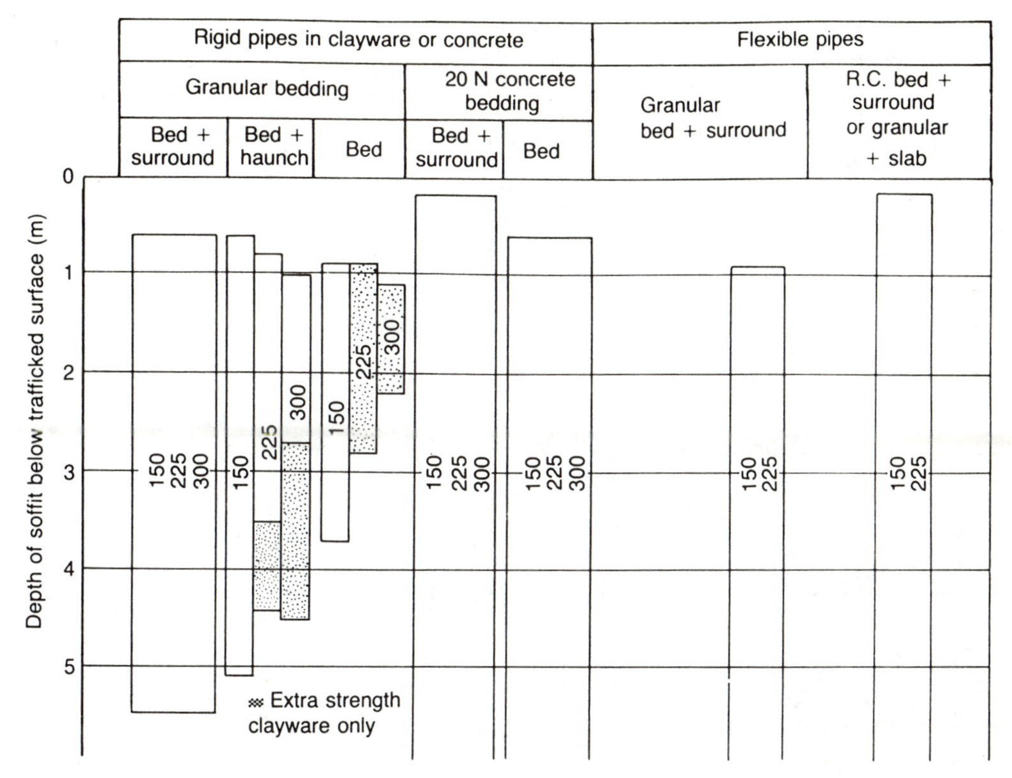

(c) Application of pipe bedding types under roads by depth and pipe diameter

underground springs. Subsurface water will generally rise to a surface within the soil, known as the water table. Subsurface water is of concern to the highway engineer in two ways.

7.4.1 Subdrains

The design of the pavement should be based on a knowledge of ground conditions including a consideration of subsurface water. It can in some circumstances be advantageous to lower the water table locally under the road, particularly to facilitate construction. In the case of a pervious pavement, such as block paving, a blanket drain may in extreme cases be necessary to remove water which has percolated through the pavement where this is likely adversely to affect the stability of the pavement or the ability of the subgrade to support it adequately. Typical arrangements for the drainage of subsurface water are shown in Fig. 7.14.

Note that the traditional 'french drain' consisting of a trench filled with coarse grained granular material is generally not satisfactory since in time it will become clogged with fines washed into it by subsurface water and, where open at the top, surface water. A woven geotextile on the external faces of the drain and a perforated carrier pipe in the bottom will prevent the entry of such particles and ensure that the drain is kept clear by the provision of a free-draining path throughout its length.

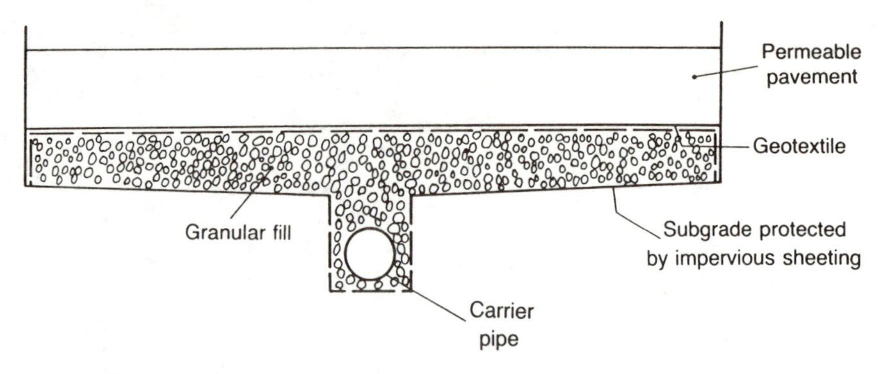

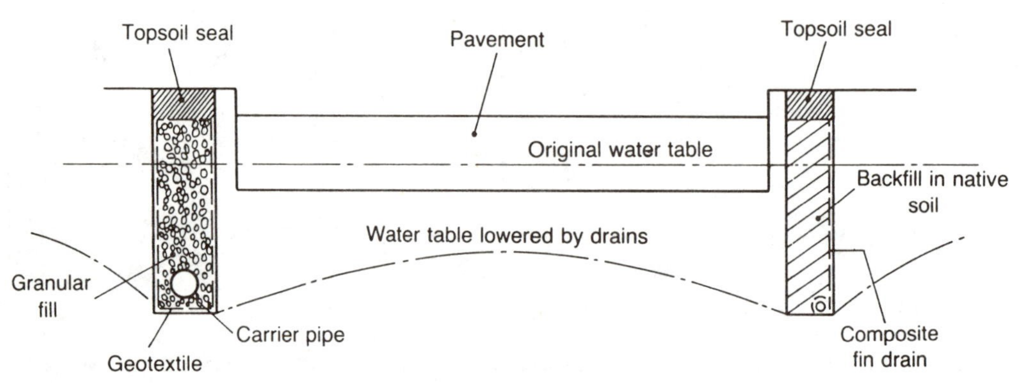

Figure 7.14 Subdrains

An alternative exists in the fin drain, which consists of a sheet of highly permeable material faced on each side with a geotextile. This system is set vertically in the soil and the lower end is connected hydraulically to a perforated plastic pipe. The geotextile acts as a filter, preventing fines from entering the system; groundwater enters through the sides of the drain, runs to the bottom and enters the pipe to be led away. This system has the advantage of allowing the excavated material to be returned to the trench after placing the fin drain.

There is no satisfactory analytical method of designing subdrains because small local variations in conditions can have relatively large effects. Instead, an empirical method is used. Typically, two trenches are dug on the positions of the proposed subdrains and these are dewatered. Boreholes or other means are used to monitor the behaviour of the groundwater between these trenches, and from these observations it is possible to estimate the correct form of the subdrains to achieve the desired effect. It is also possible to estimate the likely discharge from the subdrains from a knowledge of the discharge from the trial trenches.

Subdrains may discharge into watercourses or, if none is available, through a catchpit into the surface water drain, due allowance being made in the design. Subdrains should not discharge into foul sewers.

7.4.2 Soakaways

It sometimes happens that there is no convenient outfall to which a highway may be drained. In such circumstances the designer may, if the subsoil conditions are appropriate, choose to use a soakaway.

A soakaway provides an arrangement whereby surface water is removed by allowing it to soak into the soil. It consists of a pit into which the water is allowed to drain and through whose sides and floor the water may pass into the adjacent soil. Some means of supporting the sides of the soakaway are necessary; for some installations these may consist of perforated engineering brickwork while for larger pits an arrangement similar to those indicated in Fig. 7.14 for subdrains may be suitable. Some highway authorities are unwilling to accept soakaways as part of the public highway and so their use is often confined to private drainage systems; for the runoff from a yard and perhaps part of a roof or some such area a soakaway is a modest undertaking. For large paved areas they are often less attractive but can be of value when the occasion demands.

The ability of a soakaway to perform satisfactorily depends on its size and the nature of the soil into which it is built – clearly a free-draining granular soil lends itself much more readily to the purpose than does a heavy clay. Increasing the size of the soakaway will increase the exposed area of soil face through which water may soak into the ground, and also increases the storage capacity of the soakaway, enabling it to cope better with peak runoffs associated with short, intense storms. Escritt[7] quotes the equation

$$C = \frac{A^{1.5}}{s^{1/2}K} \qquad [7.5]$$

in which

C represents the capacity of the soakaway

A the impervious area drained

K a constant, equal to 12,400 when the units of length, area and volume are the linear, square and cubic metre respectively and the unit of time is the minute

s the rate of loss of water when the chamber is full, per unit time.

This equation is based on an approximation for the rainfall intensity.

The factor s is determined from field tests which may be conducted easily to an adequate degree of accuracy. Note that C represents the available volume of the soakaway and should be measured net of any aggregate used to fill the soakaway. An approximate allowance may be obtained by multiplying the gross volume of the soakaway by the void ratio of the material used to fill it. In the idealised case of uniform spheres this ratio will be 0.35 and for a typical graded gravel a void ratio of about 0.27 may be expected. An example will clarify this.

Let us suppose that tests show the rate of loss of water from a trial pit at a site is 0.001 $m^3/hr/m^2$ of soakaway side/floor. A car park 30 m wide is to be built at this site and will drain to a soakaway which will take the form of a trench filled with coarse filter material. This trench is to run along one edge of the car park; each metre length of the soakaway will in effect drain 30 m^2 of car park. The void ratio of the filter material is 0.3, and the water table is low. What are suitable cross-sectional dimensions for the soakaway?

Solution:
Consider unit length of the soakaway. Area to be drained, $A = 30\ m^2$.

Try depth = 1 m, width = 0.5 m: area of sides and bottom = 2.5 m^2/m run and the available volume = $0.3 \times 1 \times 0.5 = 0.15\ m^3$/m run.

In equation 7.5,

$$C = \frac{30^{3/2}}{(0.001 \times 2.5)^{1/2} \times 12{,}400}$$

$= 0.27\ m^3$; this exceeds the available volume.

Try depth = 1.5 m, width = 0.6 m: area of sides and bottom = 3.6 m^2/m run and the available volume = 0.27 m^3/m run.

Equation [8.5] gives $C = 0.22\ m^3$; this is an adequate design.

Further iterations may be necessary to achieve greater similarity between the two volumes – available and required; but it should be borne in mind that equation [7.5] is not precise.

Revision questions

1 How would you determine the best positions for gullies along a length of proposed road you have designed? Explain the considerations which lead to your conclusion and the consequences of an incorrect design.

2 Outline the information that is required to prepare a hydraulic design for a highway drainage system indicating clearly why such information is necessary. Identify where the designer needs to exercise judgement rather than simply process data and suggest how such judgement should be made.

3 Having drained water from the surface of a road, arrangements must be made for the disposal of that water. What general alternatives exist in this respect? What considerations are important in making a choice from these alternatives?

References and further reading

1 Manning, R., 'On the flow of water in open channels and pipes', *Trans. Inst. Civ. Eng. Ireland*, 1891.
2 Russam, K., 'The hydraulic efficiency and spacing of road gullies', Transport Research Laboratory, Laboratory Report 277, 1969.
3 'Design and analysis of urban storm drainage; the Wallingford procedure: Volume 4 – The Modified Rational Method', National Water Council, 1981.
4 Holland, D. J., 'Rain intensity relationships in Britain', Meteorological Office, Hydrological Memorandum No 33, 1964.
5 Bell, F. C., 'Generalised rainfall–duration–frequency relationships', *Proc. A.S.C.E.* (HY1), pp. 311–27, January, 1969.
6 'Charts for the hydraulic design of storm-drains, sewers and pipelines', Hydraulics Research Station, 1978, HMSO.
7 Escritt, L. B., 'Sewerage design and specification', *Contractor's Record*, 1947.

8. Highway design

8.1 Introduction

8.1.1 General aims of highway geometry

Previous chapters have mainly been about the materials which are used in a pavement, and the ways they can be used to build a useful road. This chapter deals with the shape of the road, and with the factors that can influence decisions made by the designer.

Different types of highway have different needs. Between the motorway and the garden path lie a range of purposes and requirements, yet all have in common the aim of providing a safe and convenient route.

In addition to the pavement engineering considerations discussed in previous chapters, the designer of a new route will seek to provide a surface whose shape and alignment is:

- useful
- affordable
- no more demanding than expected traffic can tolerate.

The way in which these requirements are balanced will depend on the type of route being provided, and the circumstances of each case.

8.1.2 The roads hierarchy

It is convenient in this context to think of roads as falling into one of a small number of different categories, according to the purpose of the road. Since purpose and form are related, geometric standards can be defined for each category.

Figure 8.1 shows a notional highway layout with roads of various types:

- principal roads are intended primarily to enable traffic to move efficiently over relatively long distances. Ideally there is little or no frontage access to principal roads, because accesses may interfere with the fast and efficient flow of traffic.

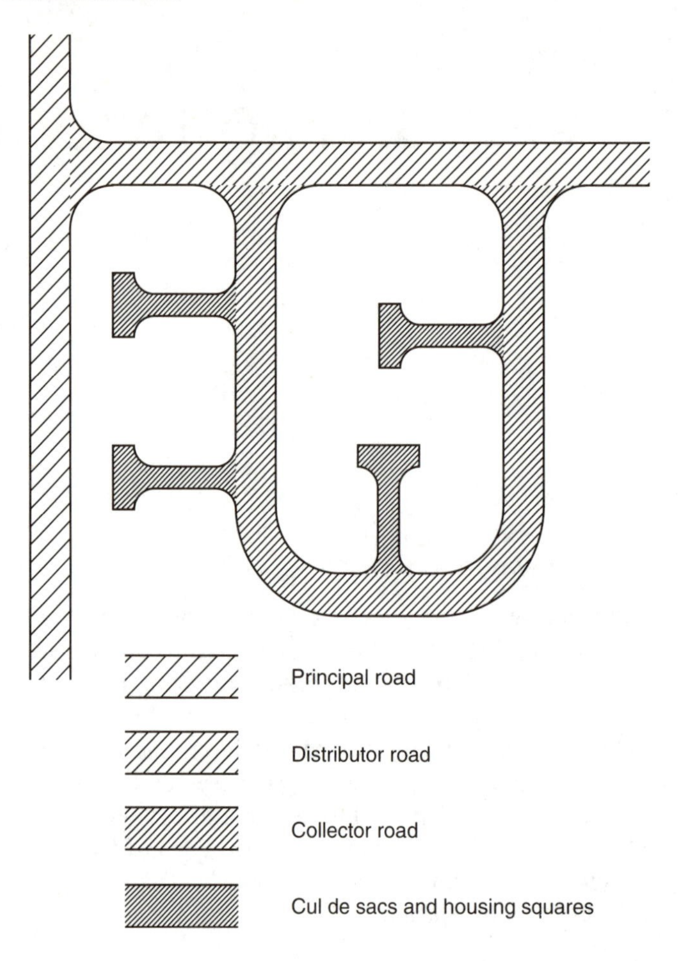

Figure 8.1 A notional layout indicating the roads hierarchy

- distributor roads are intended to allow traffic to move between principal roads and the localities where trips start and finish. There is only limited frontage access. Vehicle speeds are often lower than on principal roads.
- collector roads allow traffic to move within a neighbourhood and provide some access to individual properties. Traffic speeds are intended to be lower than on distributor roads – typically 30 mph (50 kph) in the UK.
- access roads are exclusively intended to provide access to individual properties. Design speeds are low here to allow for the large proportion of turning traffic and the frequent proximity to pedestrians.

8.2 Major roads

8.2.1 Purpose and form

Major roads are provided primarily for large volumes of traffic making journeys of more than local length. Journey times are important in such cases, particularly because the reason for building such roads is often to improve the accessibility of places along the route, and so care is taken to allow traffic to flow freely along the major road. Because traffic using major roads usually includes a higher than average proportion of large, slow vehicles it is particularly worthwhile to make provision for overtaking.

These ideas lead to major roads ideally having these characteristics:

- gentle gradients linked by long vertical curves
- long straight sections linked by curves of large radii
- junctions spaced well apart
- good standards of forward visibility
- sufficient width for traffic to pass and manoeuvre at speed
- a route determined by strategic transportation needs, major topographical features and local environmental considerations.

The last of these is not considered here.

To achieve country-wide consistency of quality, details of major road design are the subject of national standards. These are published in the UK by the Department of Transport and by the appropriate agencies elsewhere. Each standard offers a statement of best practice in each national context – standards are often a compromise between engineering, operational cost and environmental factors and there is often no universally 'right' answer. For example, in choosing the steepest acceptable gradient on a motorway, the demands of traffic suggest a relatively gentle slope but cost and environmental concerns may point to a much steeper 'maximum' with its smaller demand for earthworks.

8.2.2 Design speed

With few exceptions, the form of a major road will depend on the speed at which the designer expects traffic to move along the road. This is because fast-moving traffic needs more stopping space, and curves of larger radius, than does slow traffic, for safe and comfortable driving.

In the case of minor changes to an existing road, it is of course possible to conduct a speed survey. Traffic speeds are often 'random' (in the statistical sense) and the 85th percentile speed – that which is exceeded by only 15 per cent of all traffic (see Fig. 8.2) – is often used as the basis of design. Where changes likely to affect the speed of traffic are proposed, or where a new road is planned, a prediction method is needed.

Designers in the UK use a model[1] which predicts 85th percentile speed from the alignment of the proposed road, its layout in terms of width, junction frequency and verge width, and its speed limit. This approach is based on the view that traffic speed is affected by the degree of constraint that the form of the road suggests to the driver, and may require one or two iterations.

8.2.3 Volume of traffic

The width of a major road, and the form of junctions along it, should depend on the amount of traffic expected.

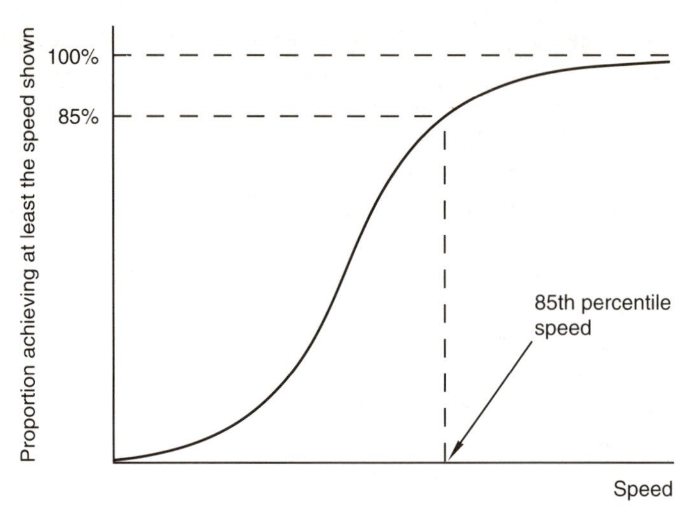

Figure 8.2 Cumulative frequency curve showing the 85th percentile speed

Prediction of traffic volumes is not an exact science. The unknowns are of two types – flow changes resulting from the new or altered road itself, and differences caused by economic variations, land use changes, and other road schemes. The general question of traffic prediction was discussed in Chapter 1, in the context of pavement design.

A development of that analysis is needed for geometric design because we are interested in the total traffic volume, rather than just the number of commercial vehicles, and because the way the traffic flow varies through the day can be important. Sufficient space should be provided for rush hour traffic, rather than merely the 24 hour average. Often the morning and evening peak hours are modelled, particularly for urban or suburban roads, although circumstances may exceptionally suggest a different approach.

In section 1.7 we saw an outline analysis of the effect of design life on the cost of a pavement, which led to a view of the optimum pavement design life. Similarly, the designer needs to take a view of the amount of time which should elapse between the opening of the new road and the time when it no longer has sufficient capacity for the traffic using it. Traffic is growing, and a road of merely adequate capacity at opening will soon become congested. Yet the provision of reserve capacity usually implies a larger road, with increased land acquisition and construction costs. It may also be difficult to justify taking land to provide capacity which may not be needed for many years. A compromise is needed.

The nature of that compromise will depend on the scheme's context. For example, many major road schemes in the UK are designed to provide for 15 years' traffic growth after opening before capacity problems are expected to arise.

Typical[2] road widths to cater for various traffic flows are indicated in Table 8.1.

Often the capacity of a road system is limited by the junctions rather than by the road links between junctions. Consideration of junction design will quickly take us into a detailed study of traffic engineering, which is beyond

Table 8.1 Approximate capacities of various road types[2] showing maximum hourly flows per carriageway

	Road configuration				
Road type	*Single two lane 7.3 m*	*Wide single two lane 10 m*	*Dual two lane 2 × 7.3 m*	*Dual three lane 2 × 11 m*	*Dual three lane motorway 2 × 11 m*
Motorway					5,700
All purpose road with restricted access	2,000	3,000	3,200	4,800	
All purpose road with frequent access	1,700	2,500			

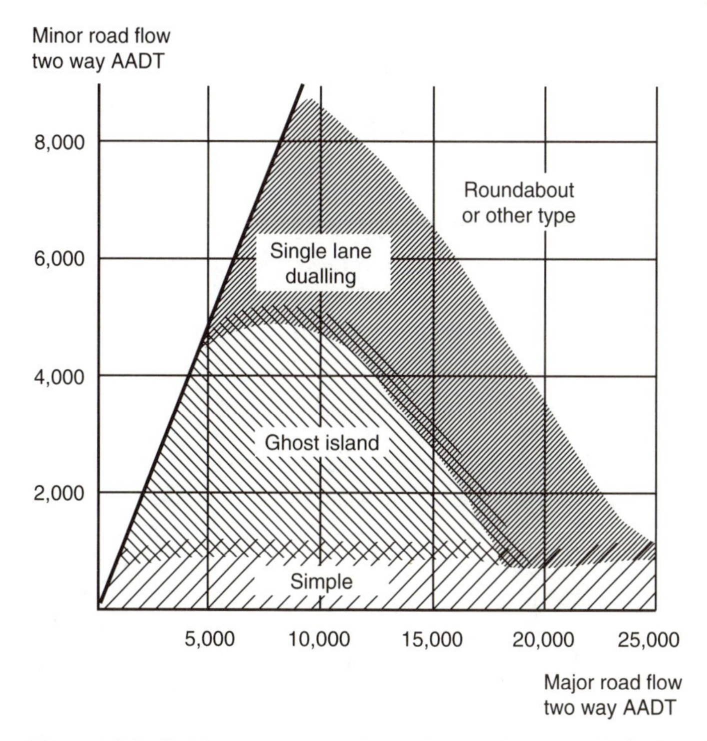

Figure 8.3 Guidance for junction type selection. Note the transition between different parts of the diagram – the designer should exercise some discretion

the scope of this book. To illustrate the principle, however, Fig. 8.3[3] suggests junction types which are likely to be best for various levels of traffic.

8.2.4 Straights and curves: the road in plan

Viewed on a plan, a road network is made up of the junctions where roads meet, and the roads themselves act as links between the junctions. The nature of these highway links should depend on the purpose and situation of the road. In any case we expect to satisfy the yardsticks of usefulness, affordability and fitness for expected traffic. In the particular case of a major road, we expect a road whose shape is such that it can be safely used by traffic moving at fairly high speeds – for example, in the UK a new dual carriageway road might be designed for traffic moving at 120 kph.

The questions of surface regularity and skidding resistance will arise here, and are considered in Chapter 4. It is also important that bends are not so sharp that vehicles go out of control or passengers feel uncomfortable, and to make sure that road users can see far enough ahead to be able to stop safely if the road is obstructed. These give us two design requirements: the need to provide at least a minimum radius at bends, and the need for proper forward visibility. Both depend on the speed at which we expect traffic to move.

8.2.4.1 Curve radii

Traffic on a major road will find its ability to travel round bends limited by the sharpness of the bend – that is, by its radius of curvature – either when the radius is so small that a vehicle travelling at the design speed becomes hard to control, or when the centrifugal forces acting on the people in the vehicle makes them uncomfortable.

Simple dynamics can show that the outward, radial force on a vehicle negotiating a bend of radius R is proportional to V^2/R, where V is the speed of the vehicle. This outward force is counteracted by a cyclist who leans into the bend, and can be counteracted in the case of other vehicles by the highway engineer who includes superelevation in the road design. Superelevation is the name given to the technique of tilting the road surface at bends so that vehicles lean towards the centre of the curve. The tendency to skid outwards or overturn is reduced by the counteracting forces, as shown in Fig. 8.4. Superelevation can allow the use of bends of smaller radii than would otherwise be possible. There is an upper limit to the degree of superelevation that can be accepted. If the transverse gradient is made too steep, limits of comfort or safety may be crossed by vehicles travelling much slower than the design speed.

Minimum radii are derived from considerations of comfort, safety and superelevation. Values frequently used are shown in Table 8.2, from which the benefits of superelevation can easily be seen.

Speed, radius and superelevation are related through the equation

$$S = \frac{V^2}{2.828R}$$

in which S is the superelevation, measured as a percentage

V is the design speed in kilometres per hour

R is the radius of curvature, not less than the minimum values shown in Table 8.2.

Superelevation greater than 7 per cent should not be used.

At the ends of a curve, as superelevation is introduced, care is needed to avoid large level areas which can cause drainage problems like those described in section 7.3.1.1. The risk of such problems can be reduced, and the appearance of the road improved, if the transition from straight to curve is made gradually.

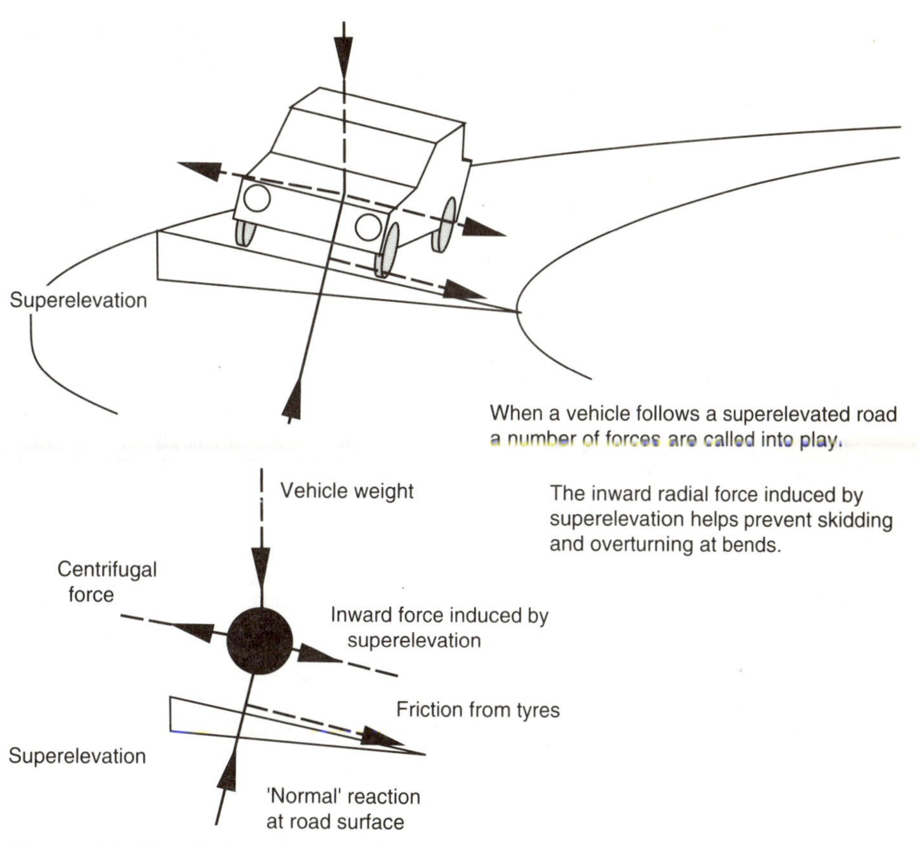

Figure 8.4 The action of superelevation

Table 8.2 Minimum radii of horizontal curves[1]

	Design speed (kph)						
	120	*100*	*85*	*70*	*60*	*50*	V^2/R
Minimum radius without elimination of adverse camber or transitions	2,880	2,040	1,440	1,020	720	510	5
Minimum radius with superelevation of 2.5%	2,040	1,440	1,020	720	510	360	7.07
Minimum radius with superelevation of 3.5%	1,440	1,020	720	510	360	255	10
Desirable minimum radius with superelevation of 5%	1,020	720	510	360	255	180	14.14
Minimum radius with superelevation of 7%	720	510	360	255	180	127	20
Limiting radius with superelevation of 7% at sites of special difficulty only	510	360	255	180	127	90	28.28

Transition curves are designed to provide a uniform rate of change of radial acceleration. The cubic spiral is often used. Its equation is

$$x = \frac{y^3}{6LR}$$

in which x is the offset from the tangent at distance y from the start of the curve
L is its length
and R is the radius at the inner end of the spiral.

8.2.4.2 Forward visibility

A driver needs proper forward visibility for three reasons:

- so that he can properly steer and control the vehicle
- so that he can see in good time any obstructions in the road
- so that he can see and make allowances for other road users.

If the last two are satisfied we may expect the first requirement to be met as well. Consequently only two standards of forward visibility are applied in highway link design:

Stopping Sight Distance (SSD) which is always needed so that a vehicle can pull up safely if the road is obstructed

and

Full Overtaking Sight Distance (FOSD) which is often provided on single carriageway roads to allow vehicles to overtake safely.

Stopping sight distance should be measured between all driver's eye positions between 1.05 and 2.00 metres above the centre of each traffic lane and all object heights between 0.26 and 2.00 metres, placed in the centre of the same traffic lane.

Full overtaking sight distance should be available between points 1.05 and 2.00 metres above the centre of the carriageway. FOSD should be provided on single carriageway major roads as often as possible – particularly where the daily traffic flows are likely to approach the capacity of the road.

Recommendations[1] for minimum SSD and FOSD are set out in Table 8.3.

Table 8.3 Minimum forward visibility – UK requirements[1]

	Design speed (kph)					
	120	*100*	*85*	*70*	*60*	*50*
FOSD (metres)	*	580	490	410	345	290
SSD (metres)						
Desirable	300	225	165	125	95	70
Abolute minimum	225	165	125	95	70	50

*No FOSD is recommended for roads in the UK with a design speed of 120 kph because the national speed limit forbids that speed on single carriageway roads.

Note that 'desirable' minimum SSD should always be achieved except when the designer has reasons for confidence that the design speed will not often be exceeded, and turning traffic – for example, from nearby junctions – is not expected. The table is based on various studies of driver reaction times and vehicle performance, and assumes a road surface of normal skid resistance and a legal tyre condition.

Forward visibility may be limited by obstacles alongside the road at bends, by the summits of hills, and sometimes by bridges or other structures under which the road sags. Checks should therefore be thorougly made in plan and in section, between all relevant points on the road.

8.2.4.3 *Visibility at junctions*

One of the essentials of junction design is to minimise the conflict that can arise between vehicles where their paths cross. Clear visibility between traffic on the major road and that wishing to enter from the side is of paramount importance, so that minor road traffic can enter the major road when it is safe, and so that major road traffic can see and anticipate the behaviour of minor road traffic. This is achieved by providing 'visibility splays' to the left and right of the minor road to provide clear intervisibility in the areas shown in Fig. 8.5.

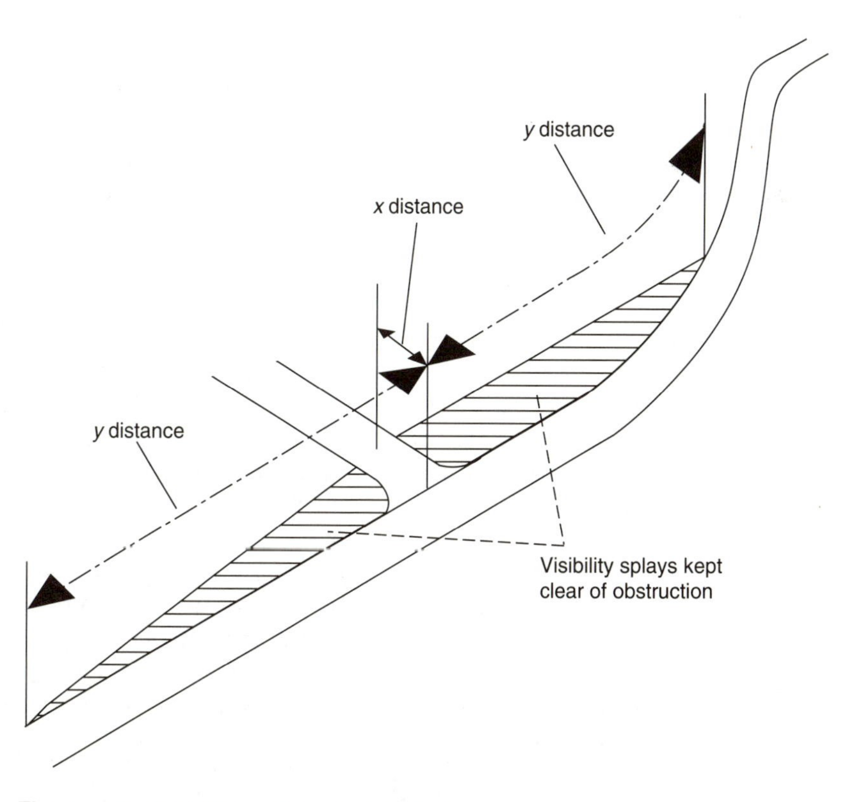

Figure 8.5 Visibility splays at junctions

In Fig. 8.5, the 'x-distance' is generally 9 metres although that may be reduced when the major road is an estate road where traffic speeds will be low. The 'y-distance' is the stopping sight distance for the major road traffic. The area inside the visibility splay should be kept clear of obstructions so that drivers in the minor road have a clear view of points 1.05 metres above the major road throughout the length of the splay. Planting, buildings, street furniture and parked vehicles can all cause problems here.

8.2.5 The road in section

The surface of a road exists in three dimensions. So far we have considered only two. The third dimension is the vertical one, often described as the 'level' of the point in question – its distance in metres above or below a reference datum. In the UK levels are often expressed in terms of their relationship to 'Newlyn Datum' which is the mean sea level at Newlyn harbour in Cornwall. The Ordnance Survey has established bench marks throughout the country whose locations and levels are shown on OS large-scale maps.

In designing a new road or analysing an existing one engineers often consider the road in cross-section and in longitudinal section. Generally, cross-sections are derived from local considerations such as the need for superelevation and the requirements of the drainage system. These are discussed in sections 8.2.4.2 and 7.3.1.1.

A road's longitudinal vertical alignment is a larger scale matter. The design problem that the engineer must solve is to provide a series of levels corresponding to points at intervals along the centre of the road as it passes through terrain which is often irregular. The solution should provide a road which

- sits in the local topography in an acceptable way
- minimises the need for material to be removed from the site or brought to it
- is satisfactory for traffic to use.

The first of these leads to design parameters such as the need to meet other roads, to pass under or over other features such as railways or waterways, and to avoid environmental intrusion.

The second comes out of the desire to minimise the cost of the road. Engineers aim to produce a vertical alignment which is partly in cutting and partly on embankment, with the quantity of material removed from the cuttings balanced by that used to form the embankments. If this can be achieved there is no need to move fill material to or from the site, and cash and environmental savings are gained. The numerical methods of achieving this are described in surveying textbooks, and built into highway design software.

Finally, there is the need for the road to be convenient for traffic to use. In this context that means that the gradients should not be too steep, and that each straight gradient should be linked to the next by vertical curves which are no more demanding than the expected traffic will be able to tolerate.

8.2.5.1 Gradients

The minimum acceptable gradient along a road is zero. This will require special arrangements to be made for drainage, but poses no other problems.

Steep gradients are disadvantageous because they slow ascending traffic, and can lead to high vehicle speeds and loss of control when descending. On the other hand, in hilly country the cost of avoiding steep gradients – by cutting or tunnel – can be high. As so often in such cases, a compromise is often made between the cost of providing the road and its subsequent serviceability. A balance will be struck between the volume and nature of traffic on one hand, and the cost in cash or environmental terms of providing an 'ideal' on the other. Table 8.4 shows current UK practice.

Table 8.4 Maximum gradients on major roads[1]

Motorways	3%
Other dual carriageways	4%
Single carriageways	6%
In difficult country where these standards may be costly or damaging to achieve, they may be relaxed to:	
Motorways	4%
Other roads	8%

Note that the standards in Table 8.4 may be very difficult to achieve in towns. On long steep ascents, extra 'crawler' lanes are often provided on single or dual carriageways to allow ordinary traffic to freely pass large heavy vehicles which can only climb the hill slowly.

8.2.5.2 Vertical curves

It is in the nature of things that no road will be of constant gradient throughout its length. The designer must arrange for each gradient to be joined smoothly to the next. As sharp angles are avoided in horizontal alignment by the use of curves, so sharp angles between different gradients are avoided in vertical alignment by using vertical curves. Designers often aim to provide constant vertical acceleration for road users by using parabolic curves.

Substandard vertical curves can sometimes be seen on old bridges over railways or canals. They are characterised by poor forward visibility as one approaches the crest, and by discomfort, caused by excessive vertical acceleration, for passengers in vehicles using the bridge. A good design avoids these pitfalls by providing a vertical curve long enough to accommodate the necessary change in gradient.

Passenger comfort is said to be satisfied if the maximum vertical acceleration is 0.3 metres per second per second.

Forward visibility requirements are the same as for horizontal curves, set out in Table 8.3. At crest curves at least the absolute minimum stopping sight distance should be provided, and more if the horizontal alignment so demands.

At sag curves visibility is not usually a problem providing that no obstacle such as an overbridge exists.

The first stage in designing a vertical curve is to make sure that it is long enough. This can be done from the following simple relationship:

$$\text{Curve length (metres)} = (\text{Difference in gradient at curve ends}) \times K$$

where the gradients are expressed as percentages, recognising the direction of each gradient. Figure 8.6 illustrates this. Values of K are shown in Table 8.5.

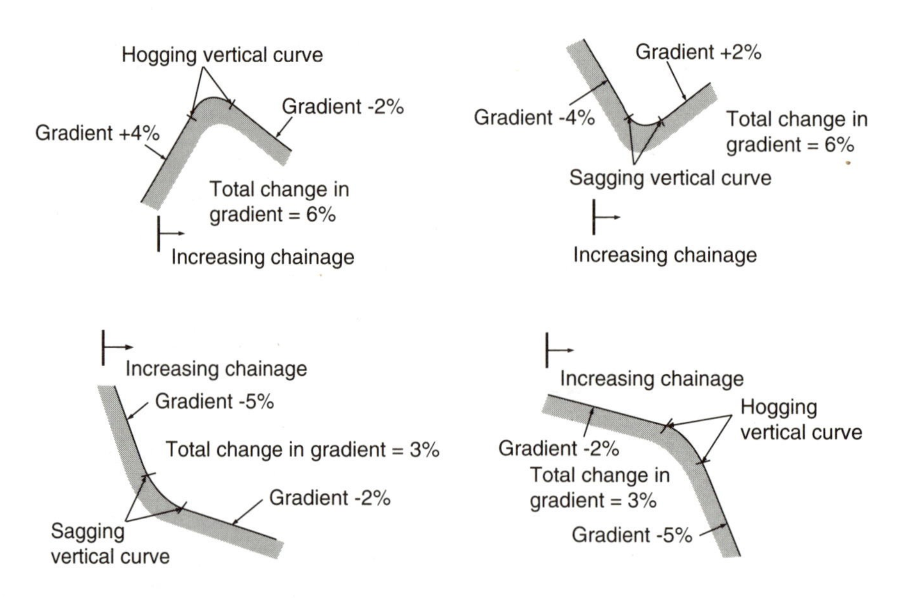

Figure 8.6 Examples of vertical curves

Suppose for example that a gradient of +2 per cent (rising in the direction of increasing chainage) is to be connected to one of −3 per cent (falling with increasing chainage). Clearly the difference in gradient at the curve ends is 5 per cent. For a design speed of 85 kph, Table 8.5 gives a desirable minimum K value of 55. Thus the desirable minimum vertical curve length in this example is 55×5 or 275 metres.

Table 8.5 Values of the vertical curve design factor *K*[1]

	Design speed (kph)					
Designing for	*120*	*100*	*85*	*70*	*60*	*50*
FOSD crest	N/A	400	285	200	142	100
Desirable minimum crest	182	100	55	30	17	10
Absolute minimum crest	100	55	30	17	10	6.5
Absolute minimum sag	37	26	20	20	20	13.5

8.3 Estate road layout

8.3.1 Introduction

The previous section of this chapter is concerned with major roads, whose purpose is to allow traffic to move from place to place safely and reasonably quickly. On arrival at the neighbourhood which is his destination, the driver's needs change to a road system which gives easy access to each site or building in the area. Having arrived, the need again changes to a desire to park and, once parked, pedestrian access and the general quality of the environment become most important. A successful estate road layout is one which properly meets the often conflicting requirements of

- access for vehicles
- parking
- access for pedestrians
- appropriate environmental quality

The way this is achieved should depend on the way the area is to be used. Two common land uses for new developments are housing and industry.

8.3.2 Housing

The roads hierarchy is illustrated in Fig. 8.1. We are concerned here with roads ranked lower than local distributors – that is, with residential access roads. The purpose of these in transportation terms is to link parking areas, serving individual dwellings, with local distributor roads. They are also an important part of the domestic environment.

This domestic element is an important one which gives residential access roads their particular character. Not only are they an important element of the setting of the homes they serve, but they also provide a part of the resident's extended living space. More than in any other case, the needs of vehicular traffic should be tempered by a recognition of the informal use of the same space by pedestrians.

8.3.2.1 Access for vehicles

Because they are not intended primarily to allow traffic to move rapidly from place to place, residential access roads have reduced geometric standards compared with those for major roads. These reduced standards are derived from the particular vehicle types expected to use the road and their needs when driven at low speed. Often the road will include arrangements intended to encourage low vehicle speeds.

From this it follows that there will be many possible options for the highway geometry of each part of a new development, and care is needed when choosing suitable standards. To overcome that difficulty, most highway authorities in the UK publish specific design guidance setting out what they feel to be best for their local circumstances. Wherever possible the designer should refer to local advice. General guidance can be given, as is set out below.

There are at least three types of residential access road:

- the collector road
- the cul de sac
- the housing square.

Collector roads provide access to individual properties and can also connect local distributors and roads lower in the hierarchy. Traffic flows will usually be more than insignificant and so footways should be provided. The geometry should not allow vehicle speeds of more than 30–40 kph. Up to about 400 dwellings can be served by a collector road in the form of a loop. Above that number, the road width may not be enough to cater for traffic generated by the development. Highway authorities often require an upper limit of 150–200 dwellings served by a road with only a single point of access, because of the increasing risk that emergency vehicles will be obstructed at the sole entrance.

Cul de sacs are roads with one access point of more or less conventional layout, giving direct access to up to about 100 dwellings. Footways should be provided, but since pedestrians and vehicles are likely to mix in the road its alignment should again constrain traffic speed. Where the road is not looped, a turning head should be provided able to accommodate large delivery and service vehicles. Figure 8.7 shows some arrangements which are often acceptable.

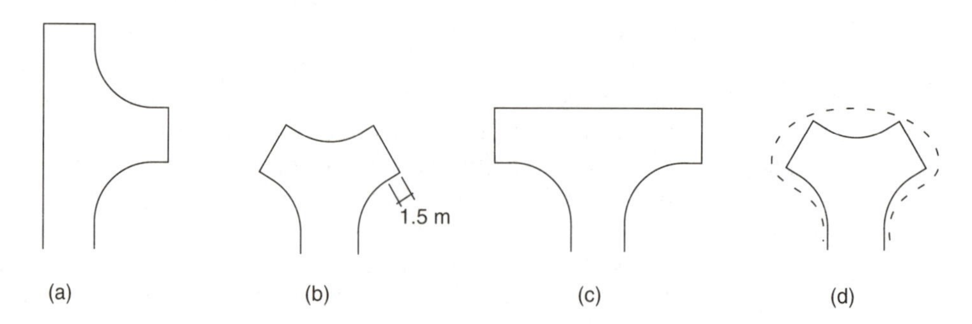

Figure 8.7 Some turning head arrangements for residential roads. Note: carriageway width is 5.5 metres; all radii are 6 metres; (d) shows how an amorphous shape may be used which includes a 'standard' layout

Housing squares are courtyards onto which up to about 20 dwellings open. They are designed as mixed use area, where pedestrians and vehicles should have equal priority. No footways are provided, but very positive measures should be taken to constrain traffic speeds to about 20–25 kph. The layout may be formal or informal and should include provision for visitor and service parking and, if appropriate, for residential parking. One of the turning areas indicated in Fig. 8.7 should also be provided at the end of the housing square.

As regards the width of the road area, this is determined not by considerations of capacity under free flow conditions, as was the case for major roads, but rather by the space needed for vehicles to use the road in the low flow, low speed conditions which the design should encourage. Carriageway width constraint can be a useful tool in controlling traffic speed. There are other ways.

For two large vehicles such as delivery vehicles or dustcarts to pass, a width of 5.5 metres is satisfactory. A large vehicle and a wide car need about 4.8 metres to pass, while two cars or a large vehicle and a cyclist need about 4.1 metres. The absolute minimum width of 3 metres will allow use by a large vehicle alone. All these widths should be increased at sharp bends to allow the widening of vehicle swept paths when cornering in cases where there is a reasonable probability of the design combination of vehicles meeting at the bend. Table 8.6 shows recommended widths for three types of access road.

Table 8.6 Recommended widths of residential access roads

Collector road	5.5 metres
Cul de sac	4.8 metres
Housing square	3.2 metres minimum, often more

Vehicle speeds can also be constrained by 'traffic calming' devices such as speed humps, speed tables, chicanes and constrictions in the road. The use of these is governed by national standards to which the reader should refer.[4] The general principles are that traffic calming should be used to prevent traffic speed rising rather than try to make it fall – in other words, traffic should enter a calmed area at a speed which has been reduced by a feature such as a junction or a sharp bend which forms part of the 'uncalmed' road system. Traffic calming should not be introduced where traffic will be travelling at such a speed that the calming features will become dangerous.

Secondly, in residential areas where traffic flows will be light, arrangements such as constrictions which rely for their effect on vehicles being forced to give way to oncoming traffic should not be used, simply because in many cases there will be no oncoming traffic and so the device will have no effect.

8.3.2.2 Residential parking

In designing any residential access road, the way in which parking will be provided should be clearly understood. If no provision is made then on-street parking – or, worse, footway parking – may reach a level at which traffic movement is impeded and the design will fail. This may also affect the ease with which the new development can be sold.

Parking is needed for residents, for visitors and for service vehicles such as dustcarts and removal vans. Usually, highway authorities specify their requirements which will vary according to house type (on the basis that car ownership depends on similar factors to choice of dwelling) and according to location (on the basis that factors such as the general availability of public transport influence car ownership). Typical requirements for parking space provision are shown in Table 8.7.

Residential parking layout should meet several requirements. It should be convenient for access on foot and by car, it should be close to and visible from the properties it is intended to serve, it should be laid out to adequate dimensions (see Fig. 8.8) and it should discourage on-street parking that will cause obstruction. This last point is illustrated in Fig. 8.9.

Table 8.7 Typical minimum parking requirements for residential areas

	Parking spaces per dwelling	
Dwelling type	*Metropolitan*	*Provincial*
4 or more bedrooms	2.1*	3
3 or less bedrooms	1.1*	2
Social housing with 2 bedrooms or less and communal parking	0.6*	1.5
Retirement dwellings	0.5*	1.25**

Notes
* Includes 0.1 visitor space per dwelling
** Includes 0.25 visitor space per dwelling

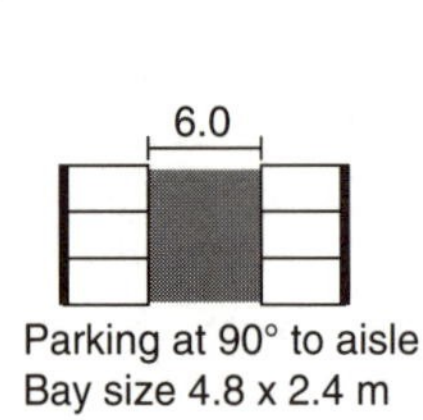

Parking at 90° to aisle
Bay size 4.8 x 2.4 m

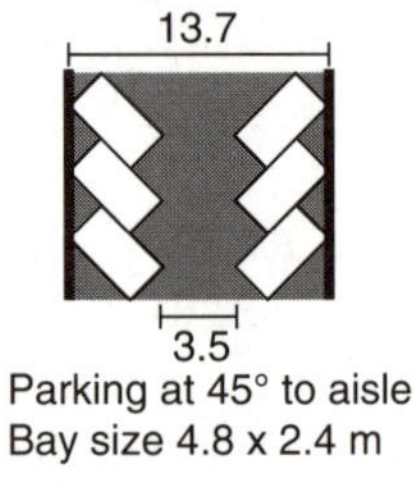

Parking at 45° to aisle
Bay size 4.8 x 2.4 m

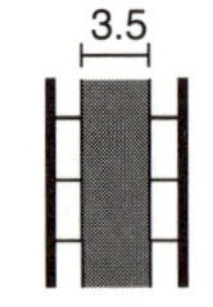

Parking parallel to aisle
Bay size 5.5 x 1.9 m

Figure 8.8 Parking bay layouts suitable for cars

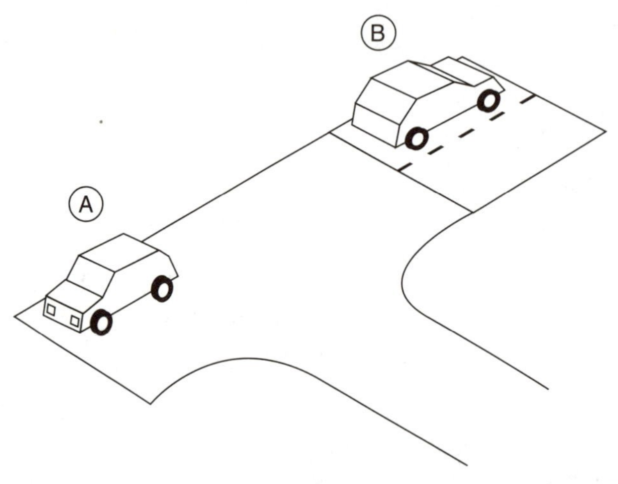

Figure 8.9 Using parking bays to prevent obstruction. The car parked at A hinders turning at the end of this cul de sac. Placing parking bays next to the turning area as at B, not only provides an alternative, but also deters adjacent parking in the turning area since to do so would more obviously obstruct other vehicles. Accesses to private drives often have a similar effect

8.3.2.3 Access for pedestrians

Pedestrian access must be universal – for all pedestrians, to all properties. This is achieved by providing a system of footways and footpaths.

Footways are routes provided for pedestrians and associated with carriageways, to which they will either be adjacent or related by a verge. Footpaths are pedestrian routes away from traffic routes. Similar standards apply to both. The distinction is drawn for administrative reasons arising out of UK highway legislation.

Generally, footpaths and footways should

- be positioned to satisfy pedestrian desire lines
- be aligned and lit in a way which will provide good personal security, for example by allowing ample visibility of and along the path
- be as level as practicable, longitudinally and transversely. Long gradients steeper than about 8 per cent are difficult for some wheelchair users
- be separated from carriageway areas by a kerb or wide verge, or be part of a mixed use area
- include no steps, which if included would make use difficult or impossible for wheelchair users, parents with pushchairs, and the like. Dropped kerbs should be provided at road crossings
- direct pedestrians to cross roads at safe and convenient points.

As with carriageways, footway widths should be sufficient for the uses to which they will be put. A lone pedestrian needs a width of about 750 mm. Two need about a metre to pass in comfort. A parent with pushchair and walking child needs about 1250 mm, while 1800 mm is wide enough for two wheelchairs to pass. Sometimes a footway is used to provide emergency vehicular access to a building, perhaps for a fire tender, in which case a width of 3 metres is often required by the fire authority. Note that all these are widths clear of obstructions such as street furniture. These translate to the guidance shown in Table 8.8.

Table 8.8 Indicative widths for pedestrian routes

Footway associated with roads serving more than about 50 dwellings	2 metres
Footway associated with roads serving less than about 50 dwellings	1.35 metres
Footway or footpath at entrances to schools and the like	3 metres
Footpaths	2 metres
Footpaths/tender paths	3 metres

8.3.2.4 Environmental quality

The quality of the street scene depends on the materials used, and the forms created with them. This is as true of the buildings, of gardens and of public open space as it is of the street itself, and these are beyond the scope of this book.

Chapter 6 identifies a range of paving materials. The appropriate choice in any particular job will depend on many factors:

- cost
- the market sector to which the development is aimed
- local authority requirements
- possibly the vernacular use of materials.

The engineer should be alive to the possibilities that can be created through the sensitive use of materials. Often multidisciplinary design teams are established to make the most of such opportunities.

Figure 8.10 shows an isomeric view of a typical small housing development and illustrates some of the points made in the preceding sections of this chapter.

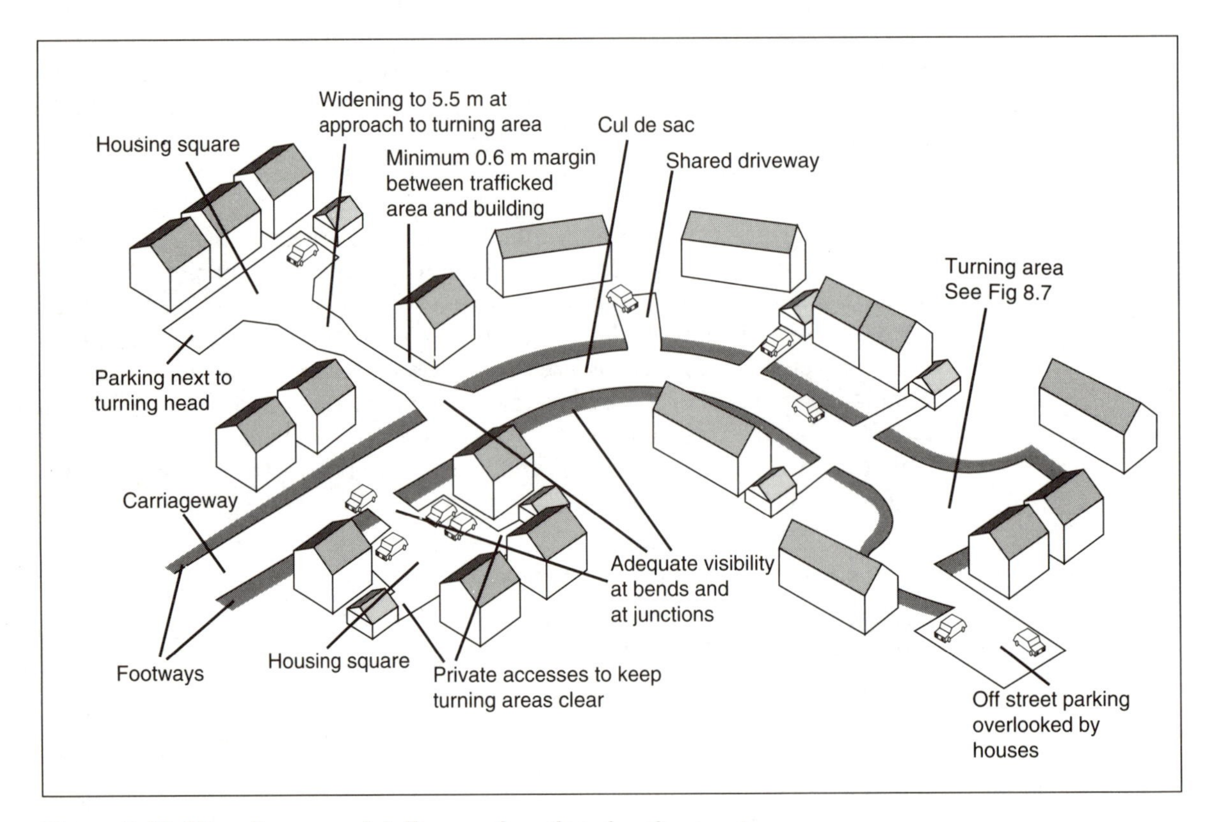

Figure 8.10 Miscellaneous details on a housing development

8.3.3 Industrial developments

Like housing estate roads, the purpose of the roads on an industrial development is to link individual properties with the principal road network. There are many differences between the two, however, because the domestic element of the environment is lost and replaced with one where vehicles of all sizes need frequent access to plots of all sizes. Although high speeds are not encouraged – because of the frequency of traffic turning movements as well as the continuing possibility of pedestrians in the road – large vehicles demand generous road space.

8.3.3.1 Access for vehicles

In this context, the concept of the roads hierarchy leads us to at least three levels of industrial road:

- the major industrial road
- the minor idustrial road
- the industrial yard or loading area.

Major industrial roads should always be considered when a development of more than about 10,000 square metres gross floor area (GFA) is to take place, and may be necessary below that level. The need will depend on the use to which the site is to be put since different types of industrial or commercial activity generate different volumes and types of traffic. It can be hazardous for large vehicles to reverse onto a busy road and so development served by major industrial roads should provide off-street parking and turning areas for large vehicles.

Minor industrial roads will always be cul de sacs and will usually serve less than about 10,000 square metres GFA. Because the level of traffic is likely to be low, more informal arrangements can be accepted for vehicle parking and turning, but this informality should not mean that adequate turning arrangements (such as those shown in Fig. 8.11) may be omitted.

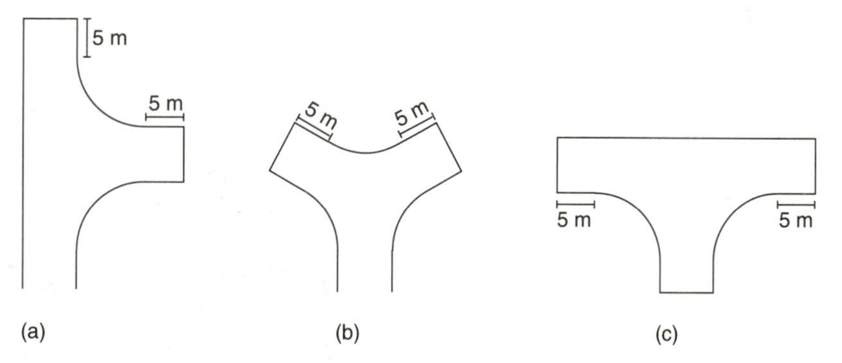

Figure 8.11 Some industrial turning head arrangements. Note: carriageway width is as shown in section 8.3.3.1; all radii are 9 metres

Generally, major and minor industrial roads should be designed to the same geometric standards as major roads. The design speed will usually be 50 kph. Major industrial roads should be at least 7.3 metres wide; minor road width may be reduced to 6.5 metres. Footways 2 metres wide should be provided on each side of the road.

Industrial yards and loading areas will not usually be part of the public highway but are nevertheless important. They arise in many contexts, from delivery bays at large supermarkets and factories to communal yards serving a number of small workshops. In each case the essential need is that they should be large enough for vehicles using the yard to manoeuvre safely and conveniently into their required positions. Each site will have its particular

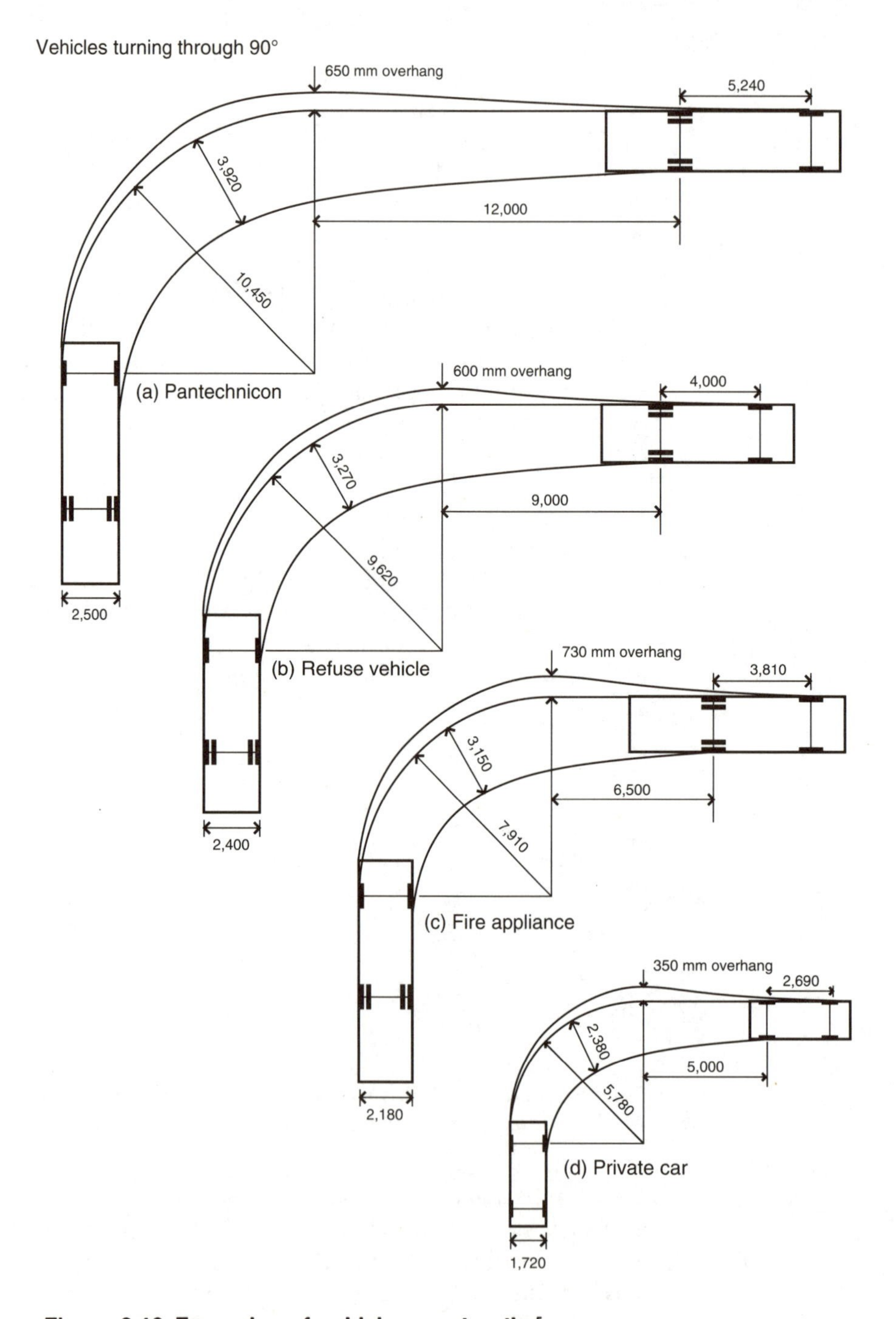

Figure 8.12 Examples of vehicle swept paths[5]

needs and characteristics and so general examples of acceptable design cannot be provided.

Instead, the approach used by most designers is to base the layout of a yard on the space needed to accommodate the swept paths generated by vehicles of the appropriate type when making the planned manoeuvre. The exact form of a swept path will depend on the particular model of vehicle to be used and the exact nature of the manoeuvre, but it is often sufficient to design on the basis of a number of 'standard' examples for which information is available. Diagrams such as those in Fig. 8.12 often give an adequate indication of what is needed. Alternatively, computer software is available to simulate vehicle movements more accuractly in specific cases.

8.3.3.2 Environmental quality

The quality of the industrial environment can be very important for some firms and so the need to meet that expectation is important to the designer in some circumstances. High quality business parks usually require access of industrial standard, for example, and in such cases the industrial access road will be in a setting of high quality landscape – and should complement that setting through its alignment and perhaps through the choice of materials for footways or carriageways. Block paving, for example, is well suited to use in industrial yards and has an appearance often considered attractive. Conversely, it may sometimes be necessary for an industrial road to do no more than meet its utilitarian purpose of allowing access.

References

1 TD 9/93 'Highway link design', Department of Transport, 1993.
2 TM H9/76 'Design flows for Urban Roads', Department of Transport, 1976.
3 TA23/81 'Junctions and Accesses: Determination of size of roundabouts and major/minor junctions', Department of Transport, 1992.
4 Highways (Traffic Calming) Regulations, HMSO, 1993.
5 Design Bulletin 32, 'Residential roads and footpaths – layout considerations', Departments of the Environment and Transport, HMSO, 1992.

9. Highway maintenance

9.1 Introduction

The highway system of a developed or developing country represents a significant part of that country's infrastructure – significant both in terms of the role it can play in the economic life of that country, and in terms of the investment made in creating the highway network. It is therefore worthwhile to expend further resources in the maintenance of the system, in order to prevent its deterioration and to enable it to properly discharge its function of allowing the easy movement of goods and people from place to place. Unfortunately, the effects of poor maintenance are not often immediately apparent – roads wear out gradually, not with catastrophic suddenness – and it is often expedient to allocate funds to areas of work which bring about more immediately perceived benefits. There are no votes in mending the road.

For this reason the engineer involved in highway maintenance is often in the position of not having enough money to treat properly all the lengths of road in his or her control. An important part of the maintenance function is therefore to be able to assess the relative demands of various lengths of road of different construction, condition and traffic density, and to determine the best allocation of scarce cash or other resources.

In the most general sense the highway includes not only the pavement but also the associated structural elements such as bridges, retaining walls, tunnels and so on, together with street furniture, lighting, drainage and many others. The general problem referred to above arises in the maintenance of these as much as in the maintenance of the pavement, but these other cases lie beyond the scope of this book. We will consider here problems which arise out of the maintenance of the pavement.

9.2 The management problem

Deciding how to allocate scarce resources is not an uncommon management problem. Often managers decide priorities and work systematically toward these. This general approach is adopted by the highway maintenance engineer, priorities being determined generally on the basis of greatest need, and objectives being clearly defined. The definition of objectives has historically been a matter of local custom and practice; this sometimes unsatisfactory state of affairs was rationalised with the introduction of common standards throughout the UK following the report of the Marshall Committee in 1970.[1] The objectives recommended by the Marshall report are summarised in Table 9.1.

Applying this rational approach across the whole highway network demands the existence of a data base drawn from a highway condition survey.

9.2.1 Preparation of a data base

The so-called 'Marshall Survey' represents the most common way of preparing an inventory of highway condition data. Traditionally, teams of technicians have toured the network, noting in a systematic way the condition of the carriageway, kerbs, footways and other elements of the highway. This information is recorded, either on forms such as that shown in Fig. 9.1 or more commonly by means of handheld computers. Further information not obtainable from simple inspection, such as deflection and skid resistance information as described below, is added to the field survey results.

The information is collated, usually by computer (a standard programme, CHART, is available for the purpose) and lengths of road identified in priority order for treatment in respect of each work area listed in Table 9.1. The maintenance standards included in the CHART suite of programmes are generally similar to those presented in the Marshall report.

Alternatively, much of the information can now be gathered from a specially equipped moving vehicle. Some examples are described in section 9.3. The practical effect of this change is that sufficient current information can be assembled to allow a realistic global view of the condition of the whole network.

Increasingly, highway authorities seek to manage this mass of data and to get the best from it by using a pavement management system (PMS). There is a cycle associated with highway maintenance management, shown in Fig. 9.2, and a successful PMS can interface with each part of the cycle. These computer based systems build on the strength of CHART to include:

- data capture
- network referencing
- condition database
- prioritisation of remedial works within overall cash limits
- lists of necessary treatments and schemes
- traffic and accident data
- management reports at various levels, from scheme- and location-specific to overviews of the whole network
- the ability to analyse data and trends for the whole network or for parts of it.

Highway managers can expect to gain from a PMS in that it should help them achieve a high level of cost effective management and planning. Priority and investment policies derived from the PMS will be soundly based, and there

Table 9.1 Summary of standards of highway maintenance

Work area	*Objective*	*Standard*
Structural maintenance of flexible pavements		
Surface dressing	Seal and bind surface, restore texture	Treat when surface is open or slippery
Patching	Repair local areas of serious deterioration	Treat as necessary
Resurfacing or Reconstruction	Correct general surface deterioration, structural failure etc. or to anticipate this	Treat when more than 30% of wheel tracks are cracked; when rutting is more than 13 mm under a 2 m straight edge etc.
Structural maintenance of rigid pavements		
Patching	Repair spalling of the surface and at joint faces	Treat when 5% of the road exhibits defect
Resurfacing or Reconstruction	Correct severe spalling, structural failure or surface irregularity	Treat when crack length exceeds 75 m per 30 m of traffic lane; when vertical movement at transverse joints exceeds 6 mm; when surface is irregular
Retexturing	Surface dressing or grooving, e.g. by saw cuts	Treat loss of skidding resistance
Joint resealing	Prevent entry of stones, grit or water	Routinely every 10 years and as necessary
Stitching	Repair of longitudinal cracking	Immediately as necessary
Other categories of highway maintenance		
Footways	Provide reasonably safe path for pedestrians at all times	Removal of trips 20 mm high or more, depressions and rocking slabs. Inspection frequency depends on location
Gulley emptying	Ensure that surface water is removed from the carriageway quickly	Width of flow in channels e.g. 600 mm max. on principal roads
Kerbs	Protect pedestrians facilitate drainage, support the pavement	Replace defective and sunken kerbs; maintain 75 mm kerbface

Note
Standards are also established in the Marshall Report for work beyond the scope of this book.

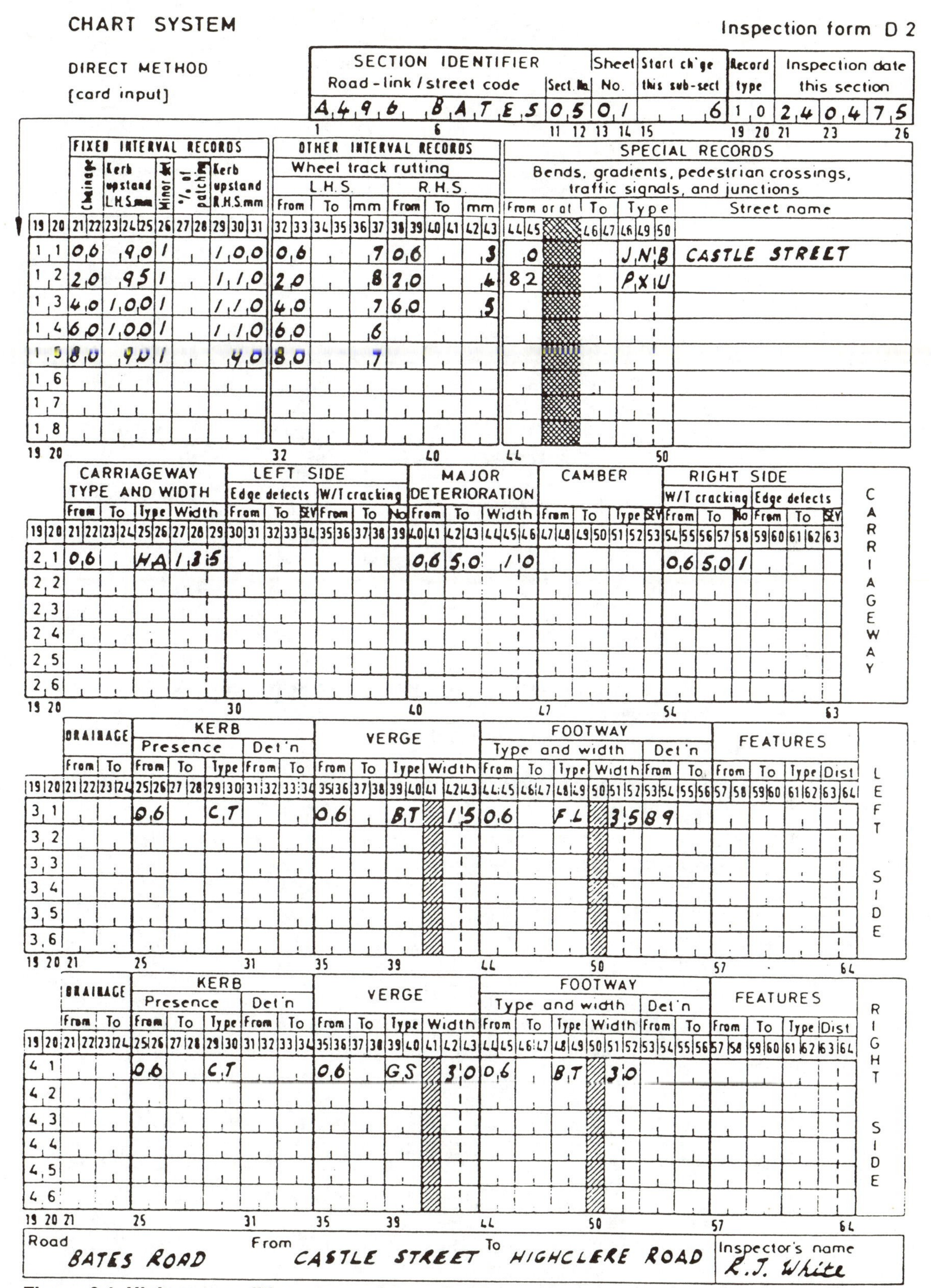

CHART SYSTEM

Inspection form D 2

DIRECT METHOD (card input)

SECTION IDENTIFIER Road-link/street code	Sect. No.	Sheet No.	Start ch'ge this sub-sect	Record type	Inspection date this section
A496 BATES	05	01	6	10	240475

FIXED INTERVAL RECORDS / OTHER INTERVAL RECORDS / SPECIAL RECORDS

19 20	Chainage	Kerb upstand L.H.S. mm	Minor def	% of patching	Kerb upstand R.H.S. mm	Wheel track rutting L.H.S. From	To	mm	R.H.S. From	To	mm	From or at	To	Type	Street name
1 1	0.6	90	1		100	0.6		7	0.6		3	0		JNB	CASTLE STREET
1 2	20	95	1		110	20		8	20		4	82		PXU	
1 3	40	100	1		110	40		7	60		5				
1 4	60	100	1		110	60		6							
1 5	80	90	1		90	80		7							
1 6															
1 7															
1 8															

CARRIAGEWAY

19 20	Carriageway type and width From	To	Type	Width	Left side Edge defects From	To	SV	W/T cracking From	To	No	Major deterioration From	To	Width	Camber From	To	Type	SV	Right side W/T cracking From	To	No	Edge defects From	To	SV
2 1	0.6		HA	135							06	50	10					06	50	1			
2 2																							
2 3																							
2 4																							
2 5																							
2 6																							

LEFT SIDE

19 20	Drainage From	To	Kerb Presence From	To	Type	Det'n From	To	Verge From	To	Type	Width	Footway Type and width From	To	Type	Width	Det'n From	To	Features From	To	Type	Dist
3 1			06		CT			06		BT	15	06		FL	35	89					
3 2																					
3 3																					
3 4																					
3 5																					
3 6																					

RIGHT SIDE

19 20	Drainage From	To	Kerb Presence From	To	Type	Det'n From	To	Verge From	To	Type	Width	Footway Type and width From	To	Type	Width	Det'n From	To	Features From	To	Type	Dist
4 1			06		CT			06		GS	30	06		BT	30						
4 2																					
4 3																					
4 4																					
4 5																					
4 6																					

Road: BATES ROAD From: CASTLE STREET To: HIGHCLERE ROAD

Inspector's name: R.J. White

Figure 9.1 Highway condition survey data collection form

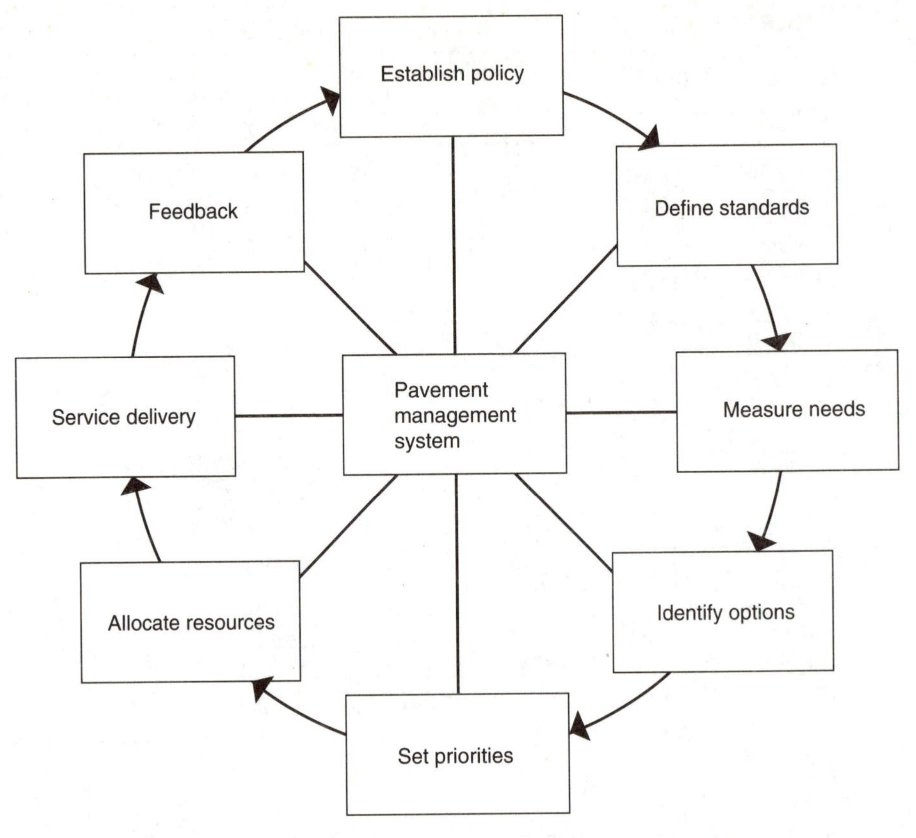

Figure 9.2 A pavement management system in the context of the maintenance management cycle

will be a sound review system to check that the outcomes of the policies match expectations.

All these things can only be achieved, however, if enough appropriate information is available.

9.3 Data capture

Apart from the simple method of looking and noting, there are a growing number of techniques, many of which are automated, available to improve both the quality and quantity of information about the condition of the highway network.

Information that can be collected in this way includes assessments of the structural condition of the pavement, its skidding resistance, and its surface texture and profile.

9.3.1 Structural condition of the pavement

The component parts of an existing pavement can be identified by trial holes or cores. The structural condition of the pavement as a whole is assessed using the deflectograph.

9.3.1.1 Cores and trial holes

Cores 150 mm in diameter may be taken at intervals of about 10 metres along the nearside wheel path in suspect areas and at larger intervals elsewhere. Information which can be obtained includes the thickness of each bound layer and the degree of bonding between those layers, the depth of cracking if cores are taken at the appropriate places, and various physical data relating to the materials. These physical data include the extent to which the binder has been stripped from the aggregate, the degree of compaction of the bound layers, the presence of detritus at interfaces between layers suggesting a loss of bond at such interfaces, and the recovery of binder samples for the determination of residual properties.

Trial pits can reveal further information, particularly in that if excavation proceeds carefully one layer at a time, the undisturbed upper surface of each layer can be scrutinised for indications of cracks, deformation and faulty workmanship. *In situ* density measurements are also possible and can be particularly valuable in the case of unbound materials, where relative differences may be identified between the wheel paths and other areas. It is also possible to investigate contamination of the bottom layer of the pavement by the subgrade. *In situ* CBR measurements of the subbase, capping and subgrade are also possible, and are to be preferred to laboratory results.

In seeking the cause of failure of an area of pavement, engineers often investigate both the failed area and a control sample of similar but unfailed pavement subjected to the same traffic elsewhere, and to seek conclusions from a comparison of results. Areas of comparison often include the composition of each layer, the penetration and penetration indices of recovered binders, the degree of compaction, the thickness of each layer and the degree of cracking found. Where no great difference is found from these investigations of the pavement itself, an obvious and generally correct conclusion is that the cause of failure lies outside the pavement – usually in the subgrade.

9.3.1.2 The deflectograph

The deflectograph is the most generally accepted tool for assessing the structural condition of a pavement.

When a load is applied to the surface of a flexible pavement that surface will deflect downwards; upon the removal of the load the surface will to a greater or lesser extent tend to assume its original form. The deflection at the time and position of application of the load is known as the maximum deflection while that which remains upon the removal of the load is the recovery deflection.

These deflections will depend on the size of the applied load, the nature of the materials used in the pavement and the thickness of the layers present, and the nature of the subgrade. In most cases most of the deflection will be elastic but sometimes, particularly when the applied load is large, some permanent deformation will take place (the recovery deflection) and it is the cumulative effect of this that causes the cracking and rutting in the wheel paths so often associated with failure of flexible pavements. All other things being constant, the permanent deflection increases with elastic, transient deflection.

The Benkelman beam is a device for measuring the deflection at a point in the surface of a pavement caused by the passage of a wheel load. It is illustrated schematically in Fig. 9.3. Deflections measured with the Benkelman beam form the basis of the empirical diagnosis of pavement defects.

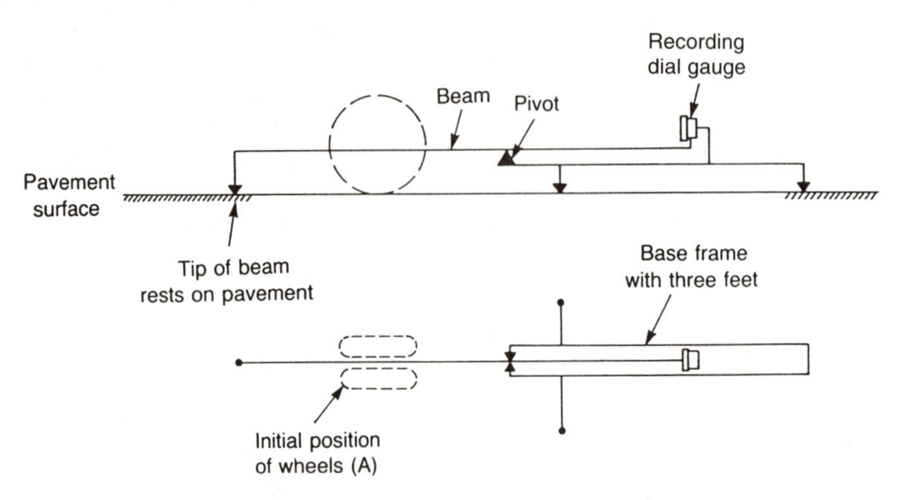

Figure 9.3 The Benkelman beam

The principle of operation of the beam is as follows. The rear axle of a two axle lorry, having two twin wheel assemblies at the rear, is loaded symmetrically to a total load of 62 kN. The rear wheels should be of a standard type and tyre pressure, with a space of about 45 mm between the walls of the paired tyres. The test starts with the loaded wheels at position A in Fig. 9.3, and the initial reading on the dial gauge is taken. The vehicle is then driven forward, past the tip of the beam and beyond so that after about 10 seconds the wheels are at least 3 metres beyond the end of the beam. During this period, the deflection indicated by the dial gauge is monitored, and the maximum recorded. The deflection at the end of the test is also noted, as this corresponds to the recovery deflection. The results of the test are temperature-dependent and so the temperature of the pavement is measured at a standard depth of 40 mm below the surface, this being facilitated by a predrilled hole filled with glycerol.

The pavement deflection is expressed in hundredths of a millimetre and is defined as being half the sum of the maximum and recovery deflection. At least two sets of results should be obtained at each point, and allowance should be made for variations in temperature above or below the standard temperature of 20 °C.

This procedure is suitable for use on short lengths of pavement – at the recommended test spacing, about 1 km of lane can be tested by an experienced team per day – but because of the time taken is less attractive for the routine monitoring of pavement condition throughout a highway network. For such purposes the deflectograph is often used.

The deflectograph operates on the same principle as the Benkelman beam, but differs in a number of important respects. Deflections are measured not by dial gauges but by electronic transducers, and are recorded automatically. Deflection measurements are made in pairs, one in each wheel track, by two beams mounted on an assembly connected to the lorry used to provide the load. This arrangement is illustrated in Fig. 9.4. The *modus operandi* of the deflectograph is that the vehicle moves forward at a steady speed of about 2 kph; in the situation shown in Fig. 9.4(a) the operating cable, connecting the deflection beam sub-assembly to the lorry, is let out at such a rate that the beam assembly remains stationary on the road. This enables the deflection caused as the rear wheels of the lorry approach the end of the beam (Fig. 9.4(b)) to be measured, after which the beam assembly is drawn forward by the cable to the position indicated in Fig. 9.4(c). The cycle is then repeated along the road, readings being obtained at intervals of about 3.8 metres.

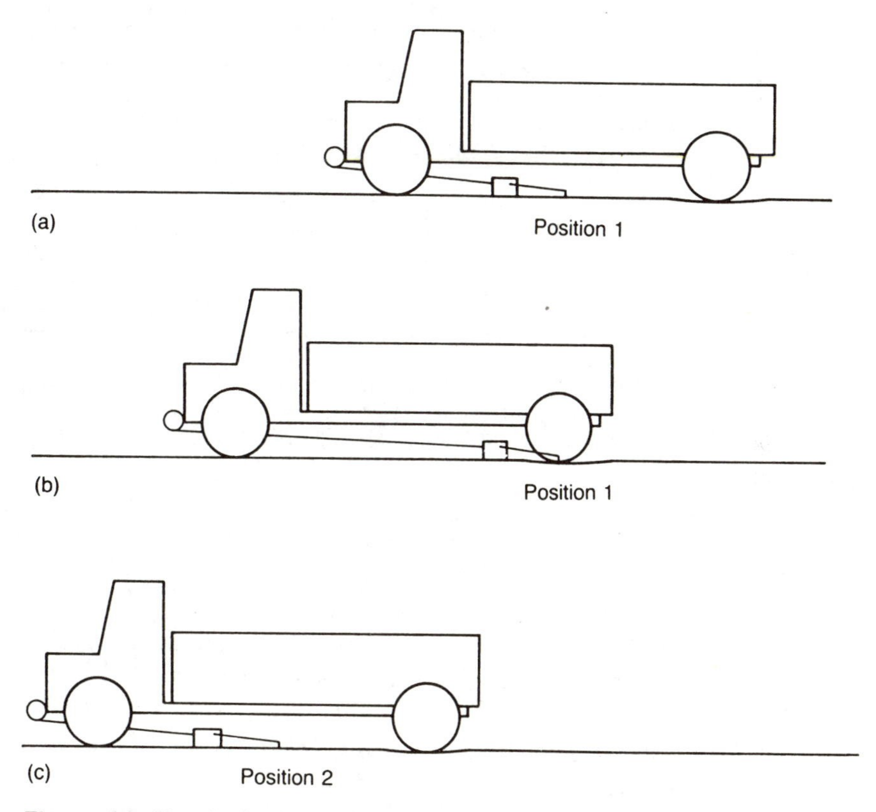

Figure 9.4 The deflectograph in use

The deflectograph gives similar but not identical results to those obtained with the Benkelman beam; operational and geometric differences necessitate the conversion of deflectograph results to enable direct comparisons to be made between the two. For various reasons the beam results are taken to be the standard.

In order that the results obtained in this way may be interpreted it is of course necessary to know which materials are present in the pavement; different materials respond to the test in different ways. The test is not applicable to rigid pavements, since these exhibit little or no deflection. Flexible pavements may be of three types – wholly bituminous, or a combination of bituminous materials and either or both of cement bound material and granular material without a natural cementing action – this latter category including for example wet mix and waterbound macadam. It is found to be sufficient to make certain generalisations about the various materials which may be found in a pavement. Guidance for the selection of an appropriate roadbase type to represent adequately any material likely to be present is given in Table 9.2.[2]

Table 9.2 Interpretation of deflectograph surveys – selection of base type

Thickness of material found in pavement construction (mm)

Cement bound		*Bituminous bound*[1]					*Granular without cementing action*						*Base type*		

1. Deduct thickness of stripped material and surface dressing layers less than or equal to 25 mm thick. Sum total of remaining layers.
2. If cement bound layer is not extensively cracked.
3. If cement bound layer is extensively cracked.

Very hard	10	1.1	*	10	1.1	*	6	1.1	1.3	6	1.2	1.5	6	1.4	1.6
Hard	14	1.0	*	14	1.1	1.5	10	1.1	1.3	6	1.1	1.3	6	1.2	1.4
Normal	20	1.0	*	14	1.0	1.4	10	1.1	1.3	10	1.1	1.3	6	1.1	1.3
Soft	#	–	–	20	*	1.3	14	1.0	1.2	14	1.1	1.3	10	1.1	1.3
Very soft	#	–	–	#	–	–	20	1.0	1.2	14	1.0	1.2	10	1.0	1.2

Notes:
1. Deduct thickness of stripped material and surface dressing layers less than or equal to 25 mm thick. Sum total of remaining layers.
2. If cement bound layer is not extensively cracked.
3. If cement bound layer is extensively cracked.

9.3.1.3 The falling weight deflectometer

The deflectograph can be criticised on a number of grounds. The load applied to the pavement acts more gradually than does that due to a vehicle moving at normal speed – this is significant because of the sensitivity of bituminous materials to load duration, as discussed in Chapter 2. The measurement of deflection at a point near the area where the load is applied does not necessarily give a full picture of the way in which the pavement responds to applied loading – the 'deflection bowl' around the load may be steep or shallow sided. The empirical origins of the design diagrams such as Figs 9.7, 9.9 and their kin renders the method of little value for pavements where little research information is available; and the method cannot be used on rigid pavements. Notwithstanding these criticisms, the deflectograph provide an adequate design tool in many commonplace circumstances. However, this is not always the case and in other circumstances a suitable alternative exists in the falling weight deflectometer (FWD).

The FWD consists in essence of a way of applying to the surface of the pavement a load whose character closely resembles that which would be imposed by a moving vehicle, and a series of geophones located at the point of application of the load and at standard distances from it whose purpose is to measure the deformations at the surface of the pavement, caused by the load. The load is applied by means of falling weights. The size of the weight and the drop may be varied, and a system of springs at the bottom of the weight's travel enables the gradual application and removal of load, characteristic of a moving vehicle, to be simulated. Vertical movements at each geophone are recorded electronically and this information, together with knowledge of the thickness of the pavement layers obtained from coring, is processed by computer to provide estimates of the strength of the various pavement layers and the subgrade, and the strain and stress distributions within the pavement. Pavement life may also be estimated, from which the residual life may be deduced by means similar to those described for the deflectograph. The principle of the FWD is illustrated in Fig. 9.5.[4]

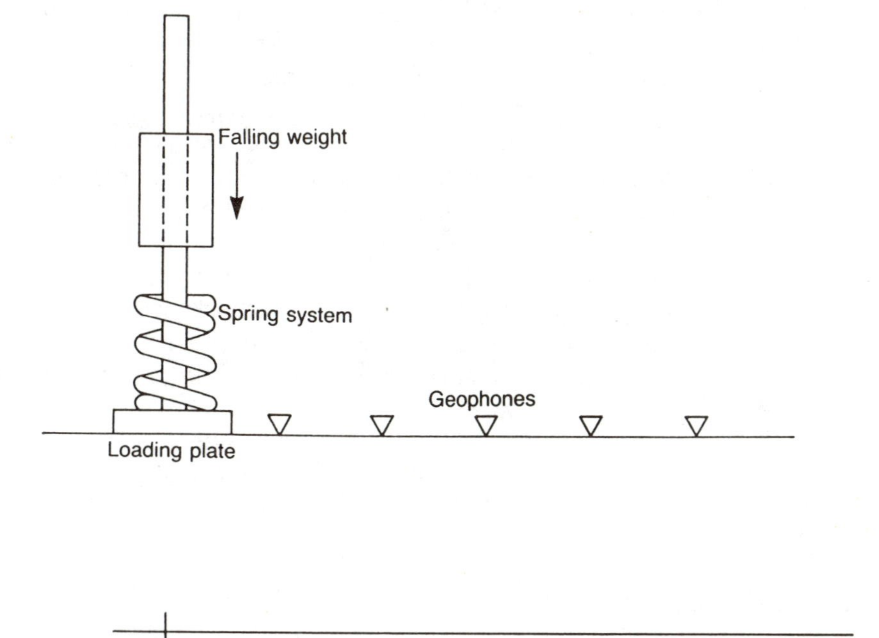

Figure 9.5 Principle of the falling weight deflectometer

From the user's viewpoint, a significant benefit of the FWD is that, since the interpretation of results is based not on empiricism but on analytical design methods, a much wider range of pavement materials can be included in the analysis, including *in situ* concrete and block paving. There is an increasing volume of relevant research work now available into the response of pavement materials to cyclical loading; in cases where this is inapplicable it is possible to carry out tests on recovered samples in order that the appropriate properties of the material in question may be determined.

No design charts are available for the FWD since the software used will generate recommended overlay designs.

9.3.2 Conditions at the surface

Here we will consider the skidding resistance of the surface, and its general profile and bumpiness.

9.3.2.1 The portable skid resistance tester

Skid resistance tests are carried out on the wet surface and are based on the coefficient of friction in such circumstances as measured by the portable skid resistance tester. This is a device of standard dimensions whose essential features are a pendulum with a standard rubber pad at the lower end, and a graduated scale. Preparation for the test consists of setting the tester up in such a way that the rubber pad bears on the surface to be tested for a standard distance at the bottom of the pendulum's swing. In the test the pendulum is raised to the horizontal and released; the pendulum falls and is slowed by friction between the surface under test and the rubber pad, but carries on of its own inertia through the plumb position and rises a certain amount before falling back. The magnitude of this first swing is assessed by a light pointer carried forward with the pendulum, which comes to rest at the peak of the swing against a graduated scale which indicates the coefficient of friction of the surface under test, expressed as a decimal fraction. The skidding resistance of the surface is reported as one hundred times this coefficient. The test should be repeated several times.

The portable skid resistance tester is also used in the measurement of the polished stone value of an aggregate (see Chapter 2 and Table 2.5 in particular).

9.3.2.2 SCRIM and the griptester

The portable skid resistance tester is slow to use, and a travelling test machine has many potential advantages. Two systems are available. Both are vehicle mounted tests which rely on measuring the forces experienced by a wheel skidding on the surface of the wet road.

In the case of SCRIM, the test wheel is mounted obliquely to the direction of motion of the vehicle. For many years this large lorry-mounted machine offered the only alternative to the pendulum, and remains the 'standard' mobile testing machine in the UK. SCRIM is an expensive machine not widely available.

The Griptester was developed for use on oil rig helidecks in the North Sea, but is now widely used on runways and public roads. It is shown in Fig. 9.6. It may be pushed by hand or towed behind a vehicle at up to 120 kph. The

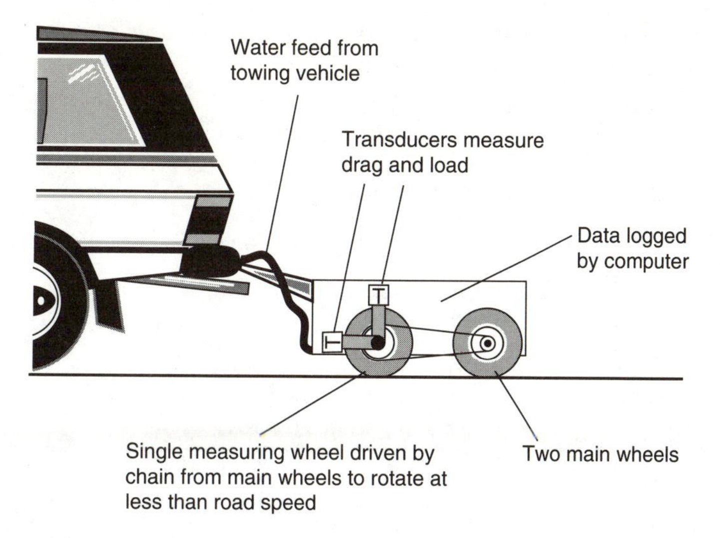

Figure 9.6 Schematic view of griptester

measuring wheel turns more slowly than the main wheels, simulating a braking vehicle wheel. Drag and load are continually monitored by a small onboard computer and location references can be entered during use so that the output can be properly interpreted. Results are consistent with SCRIM.

9.3.2.3 High speed data loggers

Hand measurements of cross-section, rutting, longitudinal profile and texture depth are usually slow and, on busy roads, difficult and disruptive. The high speed road monitor is a trailer-mounted system of laser sensors, inclinometers and distance transducers which provides continuous readout for capture in a computer on the towing vehicle. Location markers along the survey route are detected by the machine, which also measures the distance travelled from the previous marker to enable the data captured to be related to exact locations on site. The high speed road monitor can be used at speeds of up to 100 kph, and accuracy within a few millimetres is claimed.

9.4 The empirical interpretation of results

For historical reasons current practice is based on empiricism. An alternative approach is outlined later.

Common to both methods are a number of objectives which arise out of the fundamental management problem of the allocation of scarce resources. These objectives may be reduced to the treatment of areas of pavement at the appropriate time, neither too early nor too late; and the selection of appropriate techniques consistent with the cause of deterioration at each location.

9.4.1 The timing of maintenance works

A general guide to the condition of a pavement can be obtained from data obtained by inspection in the Marshall survey. A broad comparison between the condition of the surface and that of the whole pavement is given in Table 9.3.

Table 9.3 Broad relationship between pavement condition and appearance at the surface

Probable pavement condition	*Visible evidence*
Sound	No cracking. Rutting under a 2 m straight edge less than 10 mm
Critical	Either or both of: Cracking confined to a single crack or extending over less than half of the wheel track. Rutting 19 mm or less
Failed	Either or both of: Interconnected multiple cracking extending over the greater part of the width of the wheel path. Rutting 20 mm or greater

This table illustrates very clearly the sequential nature of pavement deterioration. The engineer seeks to anticipate the onset of critical conditions by the timely carrying out of maintenance work immediately before this condition is reached. Such an approach will obtain the best 'value for money'. Work carried out after critical conditions have been reached tends to be much more expensive to achieve the same increase in pavement life, while work other than superficial measures such as surface dressing carried out while the pavement is still sound will generally have less effect in extending pavement life than would be the case if they were delayed to the onset of critical conditions.

In the context of a pavement exhibiting signs of structural distress, the usual treatment is to provide an overlay or to reconstruct. It will occasionally be the case that other remedies are appropriate, and these will be indicated from the survey work outlined above; but generally these two options will be the only worthwhile ones. Specific questions which should be answered include the necessary thickness, material and time of application of the overlay. The deflection concept provides the key to answering the first and last of these; if the required thickness of overlay is unattractively expensive, or if site constraints prevent the application of the required thickness of additional material, reconstruction of the pavement should be considered.

It is generally the case that the larger the standard deflection (deflection as measured and corrected to standard test conditions) the smaller the probability of that pavement achieving a long life – its life being measured as before in terms of the number of standard axles it can support before the onset of critical conditions. If the pavement is stiffened by the addition of an overlay, its standard deflection will be reduced and the probability of its achieving long life will be increased. Figure 9.7[3] indicates the relationship between standard

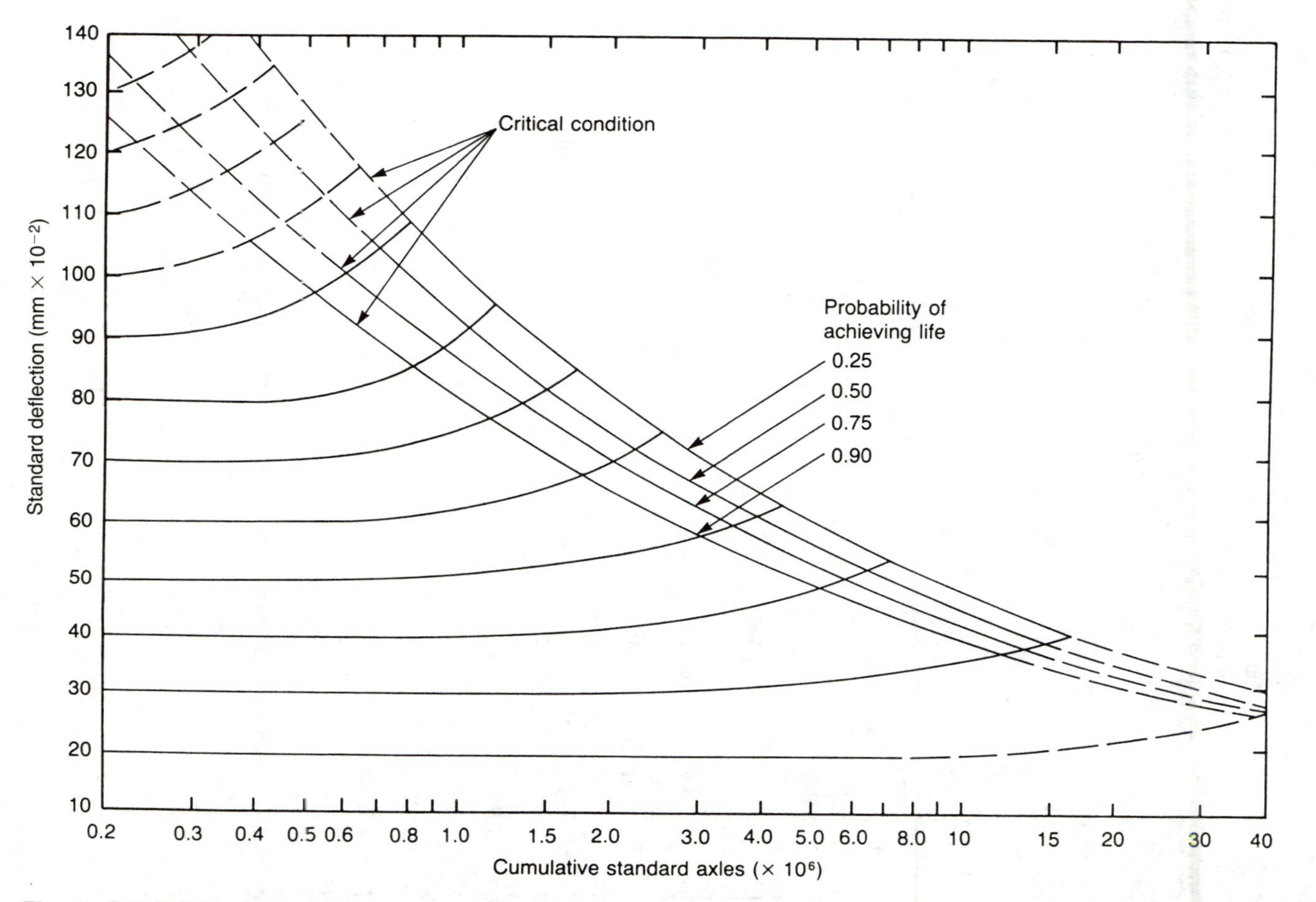

Figure 9.7 Relationship between standard deflection and life for pavements with bituminous roadbases

deflection and pavement life for wholly bituminous pavements. It should be noted that the pavement life is the whole life of the pavement since it was opened to traffic in its current form, and not an indication of the residual life at the time of the deflection survey.

It is however the residual life which is of the greatest interest to the engineer since it will be the expiry of this life that will demand the application of an overlay, or other remedial work. It is therefore worthwhile to estimate the volume of traffic – in terms of millions of standard axles – that has used the pavement to the time of the test. Figure 9.8 provides a way of estimating this traffic volume and equation [1.1] enables the appropriate vehicle damage factor to be calculated. From these the cumulative number of standard axles may be estimated, and by subtraction the remaining life of the pavement may be obtained from Fig. 9.7 or by other similar diagram appropriate to the pavement's construction.

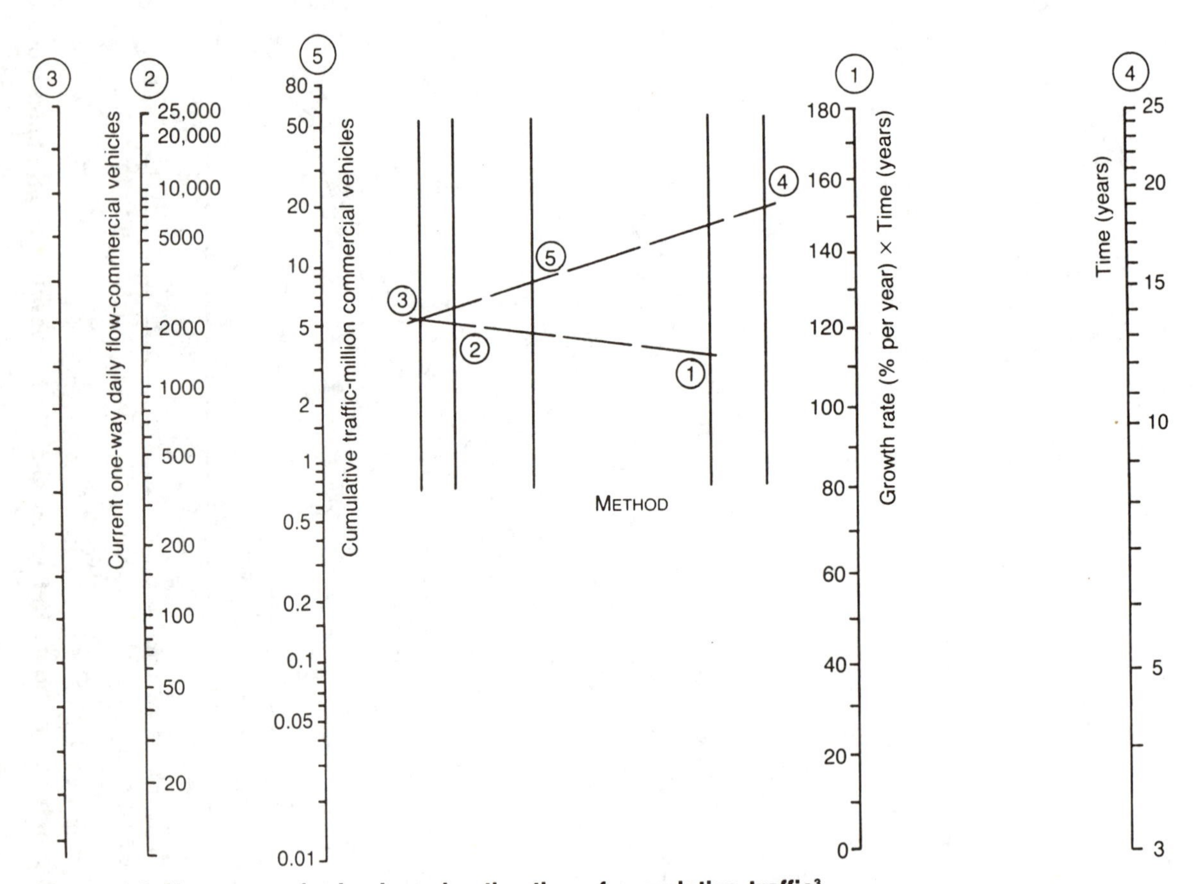

Figure 9.8 Nomogram for backward estimation of cumulative traffic[3]

An example may clarify this. Suppose the following data are available:

Deflection, 40×10^{-2} mm on one length of pavement
60×10^{-2} mm elsewhere

Age of pavement, 10 years

Historic growth rate of traffic, 3 per cent per annum

Current rate of flow of commercial vehicles, 800 per lane per day

Historic vehicle damage factor over this 10 year period, 0.91

Pavement construction, wholly bituminous roadbase and surfacing

This information may be interpreted thus:

Entering column 1 of Fig. 9.8 at 30 (% years) and extrapolating through 800 daily flow in column 2 gives an intercept on column 3. Joining this intercept to the 10-year point in column 4 passes through column 5 at a point corresponding to 2.2 million commercial vehicles (mcv) total traffic.

2.2 mcv $\times$ vehicle damage factor of 0.91 gives total traffic so far on the pavement equal to 2 msa.

In Fig. 9.7, a pavement with a bituminous road base and a standard deflection of 40×10^{-2} mm is shown to have a 0.5 probability of achieving a life of slightly more than 6.0 msa before the onset of critical conditions. This corresponds to several years more use at the current rate.

Where the deflection is 60×10^{-2} mm, the total life of the pavement is suggested by Fig. 9.7 to be little more than 2 msa; conditions in this case are such as to suggest that strengthening of the pavement is needed almost immediately.

9.4.2 Overlay design

To pursue this example a little further, the remaining question has regard to the form of the overlay – how thick should it be? A thin overlay will be relatively cheap, but will not strengthen the pavement to the same extent – and therefore achieve such a long lasting result – as would a thicker overlay. The engineer should decide what increase in pavement life is required; this objective will typically be initially expressed in years, but must be reduced to terms of a number of standard axles in order that the design can be concluded. Let us suppose that a further ten years life is required, and that the VDF is expected to be 1.1 during this period. Use of Fig. 1.7 for the forward estimation of traffic leads us to an estimate of traffic flow during the next ten years of 3 mcv, which in this case equates to 3.3 msa.

We are therefore seeking to strengthen the pavement to achieve a total life of 5.3 msa.

Figure 9.9[3] presents recommendations for overlay design. In the case in question, the appropriate thickness of overlay is indicated by the intercept of the 60×10^{-2} deflection curve with the 5.3 msa ordinate, corresponding to an overlay thickness of 50 mm at the 0.5 probability level – that is, the design will be acceptable if that level of probability of survival for the stated duration is acceptable. If a higher probability of survival is required, a greater thickness of overlay will be needed. Charts similar to Figs 9.7 and 9.9 are available for other road base materials and other levels of probability. Figure 9.9 is based on the use of rolled asphalt as the overlay material. Thicknesses obtained from the chart should be multiplied by 2.0 if open-textured macadam is to be used, by 1.3 if dense bitumen macadam with a 200 pen binder is to be used, or by unity where dense bitumen macadam with a 100 pen binder is proposed.

In cases where a thick overlay is indicated as being appropriate, or where

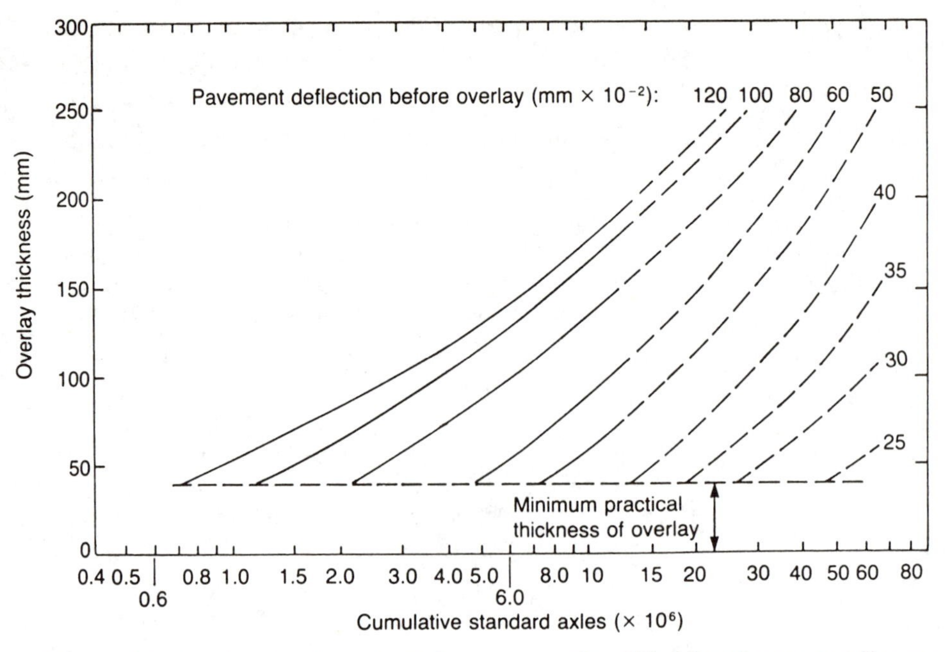

Figure 9.9 Overlay design chart for pavements with bituminous roadbases (0.50 probability)

site constraints prevent the application of an overlay of even modest thickness, the engineer may consider the reconstruction of the pavement. Alternatively, in the latter case it may be possible to remove defective surfacing materials, obtain deflections on the reduced pavement, and design an overlay on that basis – this last approach requires resources and contractual flexibility not always available.

Note that the minimum recommended thickness of the overlay is 40 mm. Experience has shown overlays thinner than this to be prone to premature failure, and to have little strengthening effect.

9.5 Repair techniques: flexible pavements

Having assessed the maintenance needs of the various lengths of road, and established priorities, the engineer should select a treatment method for each case. A number of repair techniques are available, which may be generally categorised thus:

1 the addition of a new element in the physical composition of the pavement
2 the replacement of parts of the pavement
3 the modification of the properties of the original pavement

The most commonly used techniques were mentioned in Table 9.1.

9.5.1 The addition of a new element

9.5.1.1 Resurfacing

Resurfacing should be considered when it will achieve a worthwhile increase in pavement life, or when an existing surface has deteriorated – for example by rutting or fretting – but the rest of the pavement is known to be sound.

The design of a bituminous overlay to an existing pavement should address two general problems: the material specification and the thickness of such material required. The first of these is similar to that for the flexible surfacing of new pavements, described in Chapter 4; the second has been described previously in this chapter.

The construction of an overlay should follow the same general principles as those described for any bituminous surfacing – with the proviso that when working on a road open to traffic attention should be paid to the needs of traffic passing through the site. This problem can be at least as complex as those which arise out of the construction process itself!

9.5.1.2 Surface dressing with bituminous binders

Surface dressing is usually considered worthwhile where the surface of a flexible pavement is generally open or finely crazed, or not impervious to water; where deterioration occurs over more than 10 per cent of the wheel tracks in the nearside traffic lane or 5 per cent of the whole carriageway surface; or when the skidding resistance of the surface falls significantly below the requirements indicated in Chapter 4.

Such dressing consists of applying binder to the road surface, placing chippings in the binder film, and taking any necessary steps to ensure that the chippings are firmly held in the binder. Variables which the designer must control include the nature of the binder, and its rate of application to the surface; the type and grading of the chippings and their rate of application; and the nature of any necessary treatment after the chippings have been applied. Variables which constrain the designer are at least threefold: the nature of the surface to be treated; the traffic conditions on the road; and the weather conditions at the time of doing the work. A satisfactory design can result in a surface dressing with a life in excess of six years.

These constraints influence the design in various ways. A design objective is to select the chipping size so that after the action of traffic, forcing the chipping into the original surface, sufficient macrotexture remains for the surface to function properly. Thus a hard, lightly trafficked surface would require a small chipping which on a softer, heavily trafficked road the chippings will tend to be driven deeper into the surface and should be correspondingly larger. The hardness of the original surface may be measured by a standard arbitrary test – measuring the penetration of a standard rod when applied to the surface under the action of a standard load and at a standard temperature – and categorised as being very hard, hard, normal, soft or very soft. Traffic flows are also categorised, in terms of the approximate current number of commercial vehicles per lane per day.

Table 9.4 Surface dressing: recommended nominal size of chippings (mm) and binder spread rates (litres/m²)

Type of surface	Traffic category (Commercial vehicles per lane per day) 2,000–4,000			1,000–2,000			200–1,000			20–200			0–20		
	A	B	C	A	B	C	A	B	C	A	B	C	A	B	C
Very hard	10	1.1	*	10	1.1	*	6	1.1	1.3	6	1.2	1.5	6	1.4	1.6
Hard	14	1.0	*	14	1.1	1.5	10	1.1	1.3	6	1.1	1.3	6	1.2	1.4
Normal	20	1.0	*	14	1.0	1.4	10	1.1	1.3	10	1.1	1.3	6	1.1	1.3
Soft	#	–	–	20	*	1.3	14	1.0	1.2	14	1.1	1.3	10	1.1	1.3
Very soft	#	–	–	#	–	–	20	1.0	1.2	14	1.0	1.2	10	1.0	1.2

Notes:
Column A – chipping size (mm)
Column B – rate of spread of cutback bitumen, litres/square metre
Column C – rate of spread of K1-70 emulsion, applied hot
* Binder not recommended here
# Not suitable for surface dressing
After dressing with 20 mm chippings the surface of the work must be thoroughly swept clear of all loose chippings which may otherwise cause damage to windscreens.

Appropriate chipping sizes for various traffic/surface hardness combinations are shown in Table 9.4.[5] This table also shows target rates of spread of binder.

The weather influences the surface dressing process. Cold weather will demand the use of a binder of low viscosity – in order that it will remain fluid to accept the chippings – which will result in rapid and serious loss of chippings both when the road is first trafficked, and later in warm weather. Wet weather prevents adhesion between binder and aggregate and between binder and road surface. In the UK suitable conditions are likely to be found only during the summer months. The 'high season' for surface dressing is from May to mid July, with work being possible on less heavily trafficked roads (less than about 1,000 commercial vehicles per lane per day) in the south of the country (England and Wales) from April to September. Some suitable binders are indicated in Table 9.5.[5]

Table 9.5 Surface dressing: binder viscosities for use in the UK

Traffic category (cv/l/d)	Time of year	Binder: Cutback bitumen (STV)	Spraying temp. (°C) (target)	Bitumen emulsion	Spraying temp. (°C) (target)
4,000–2,000	May–mid July	200 sec	160	–	
2,000–1,000	May–mid July	200 sec	160	K1-70	80
0–1,000	June–August	100 or 200 sec	150	A1–55, A1–60	Ambient
0–1,000	April, May September	100 or 200 sec	150	K1–60, K1–70	temp.

The prime variable which the designer can control is the nature of the chippings – particularly where surface dressing is to be used to improve the skidding resistance of the road. Care should be taken that the chippings meet the normal requirements for aggregate abrasion and resistance to polishing described in Chapter 4 for new wearing courses. Chippings may be precoated with a thin layer of binder, whose purpose is to prevent dust on the chippings and to promote rapid adhesion to the binder film at the time of laying; or they may be uncoated. The difference in cost is marginal, but despite the possibility of a slightly inferior result the uncoated chipping is sometimes specified because of the pleasing appearance it lends the road. The rate of spread of the chippings should be such as will produce a single layer of chippings completely covering the surface, with little or no excess. The actual rate will depend on the density of the stone, the nominal size of the chippings and their shape; typical ranges which are likely to be appropriate are indicated in Table 9.6.[5]

Table 9.6 Surface dressing: ranges of rate of spread of chippings

Nominal size of chippings (mm)	*Range of rate of application (kg/m²)*
6	7±1
10	10±1
14	13±2
20	17±2

The choice of binder is another significant variable. The commonest binders for general use in the UK are either cutback bitumen or bitumen emulsion. On very hard surfaces and at the higher traffic levels the elastic properties of cutbacks should be modified by including rubber or a suitable polymer. On other surfaces and where traffic is not expected to exceed 200 commercial vehicles per lane per day emulsions other than K1–70 may be used (Table 9.5 refers) and these may be applied cold in such circumstances.

Having determined the specification for surface dressing, one should ensure that work on site is carried out satisfactorily. Areas of particular interest include ensuring that the binder is applied uniformly – within 10 per cent of the target rate of spread – and only where it is needed; ironwork and the like in the road should be masked before work starts. The surface should be dry and free from dust, although emulsions may be applied to damp surfaces. Binder temperatures should be as indicated in Table 9.5. After the application of chippings the work should be rolled by pneumatic-tyred rollers which have the advantage of acting over the whole surface – unlike steel-tyred rollers which act primarily on high spots and often span over local depressions; indeed, repeated use of heavy steel rollers may cause localised damage to the work as chippings at any such high points are crushed by repeated rolling in this way.

After rolling it will generally be necessary to open the area to traffic. If traffic is allowed to drive over the work at normal speed aggregate will be lost from the surface and vehicles possibly damaged by loose chippings. Speeds must be kept down to about 30 kph (20 mph) until the surface dressing has fully stabilised by hardening of the binder. This may be done either by erecting warning signs, or more positively by heading each platoon of traffic passing over the work with a controlled vehicle driving at the required speed. Loose chippings should be properly removed from the work, both on the day of laying the dressing and as necessary thereafter.

9.5.1.3 Surface dressing with other binders

Although the use of bituminous binder can give several years' service at most sites, where there are substantial turning movements or where vehicle braking forces are large this will not be the case. An alternative which is much better able to survive repeated horizontal forces is surface dressing with an epoxy resin binder. Because of the ability of this system to resist such forces it has found application at particularly high risk sites such as the approaches to pedestrian crossings, heavily trafficked junctions and the like. To further enhance the performance of the material in such circumstances the aggregate used is generally small sized calcined bauxite of high PSV.

This resin based highly skid resistant surface treatment differs from conventional surface dressing in two other ways. In order that the excellent strength properties can best be exploited it is important that the surface on to which the material is sprayed is itself sound and clean. The cost of the system is also significantly higher than that of conventional surface dressing – typically the more sophisticated process is perhaps ten times more expensive than the bitumen-based system. This high cost has to some extent constrained the use of the system and, since most of the extra cost is in the binder, the sensitivity of the necessary rate of spread of the binder to the rugousity and general condition of the substrate can be a significant factor when costs are being calculated. A typical specification requirement is that the rate of spread of the binder, in conjunction with a 3 mm nominal size aggregate, should be not less than 1.35 kg/m^2.

9.5.1.4 Slurry sealing

Surface dressing is sometimes seen as being unattractive by virtue of the need for care when the surface is first opened to traffic, the need to remove loose chippings, and the possible consequences of failing to do so. Other means of applying the binder/aggregate system were therefore investigated, and a result of these investigations is slurry sealing.

The slurry used in this process is a mixture of fine aggregate, graded from about 3 mm down, and a bitumen emulsion whose particular characteristic is that it should be fast breaking so that the slurry should stabilise quickly. The emulsion should make up about 20–25 per cent of the weight of the slurry. The aggregate is usually crushed rock. This mixture is agitated in the laying machine to a creamy consistency and then allowed to flow onto the road surface. An additive in the form of Portland cement is often included in the mix to control its consistency and prevent segregation. Typically the laying machine advances over the surface at a brisk walking pace. It may be necessary to provide a tack coat in cationic emulsion before slurry sealing, in order to promote adhesion.

Because of the small sized aggregate used, the macrotexture achieved with slurry sealing is not such as will impart more than an average skidding resistance to the road and for this reason the technique is not used on major routes. However, it is of value in treating local roads, particularly in cases where the surface of the road is starting to craze and the sealing of cracks is an objective.

The system is priced comparably with surface dressing, but its use is not confined to the summer months. Slurry sealing can be undertaken in any circumstances where the use of a bituminous emulsion is not inconsistent with good practice – where the temperature is above freezing and in dry or slightly damp conditions. Traffic may be admitted to the work as soon as it has stabilised; the actual time for which will depend on the nature of the binder and the ambient conditions but which can be as little as a few minutes.

9.5.2 Replacing parts of the pavement

If the analysis of a pavement leads to the conclusion that the addition of material at the surface will not provide the desired result in the best way, the designer should explore other approaches. Replacement of part or all of the pavement is one such approach.

When the cause of failure lies in the subgrade or subbase, or in the fundamental inability of the pavement to protect the subgrade from the action of traffic, the removal and replacement of the complete pavement may be the most attractive option. Such reconstruction should be designed in the same way as would a new pavement, although possible constraints on the design process may include the need to keep the road open for traffic, and to make allowance for the presence of underground services beneath the road which may impose limits on the depth of construction – especially if these services are vulnerable to damage from construction plant working on the new formation.

Elsewhere, it may be possible to achieve the required objective by the partial reconstruction of the pavement – perhaps confining work to the bound layers of a flexible pavement, with work to the subbase limited to reshaping and recompaction as necessary after removal of the roadbase. It may also be worthwhile to limit the area of reconstruction to only a part of the pavement – perhaps only a single traffic lane may require treatment.

9.5.3 Modifying the properties of the original pavement – recycling

A layer of failed bituminous material has in itself a certain value in that the aggregate and binder present would cost a significant sum to replace. Using conventional overlay or reconstruction techniques this potential value is at best not realised and at worst reduced to a liability by the cost of its removal. The recycling of flexible paving offers a way of making use of these valuable materials.

Recycling techniques of various types have been in use throughout the world for some years and have established themselves as methods of producing satisfactory results at lower cost than resurfacing or reconstruction. Recycling techniques have in common the mixing of a stabilising agent with coated roadstone salvaged from the original pavement to form a material which can be successfully re-used. The techniques differ in the choice of stabilising material, the method of mixing, and the need for after-treatment before opening to traffic.

9.5.3.1 *In situ* stabilisation

The key to the *in situ* mixing of roadstone with a stabilising agent has been the increased availability of suitable plant. The various types available are essentially rotovators as mentioned in Chapter 3 in the context of soil stabilisation, but are much more powerful than those originally designed for agricultural use. A typical stabilising machine has tungsten tipped tines, powered by a 225 kW motor and turning at perhaps 250 rpm, and is capable of pulverising bituminous pavements up to 300 mm deep, working in strips 3 metres wide and advancing at up to 12 metres per minute. The machines include a mixing chamber within which the tines rotate, where the roadstone may be mixed with the stabilising agent.

Stabilisation may be achieved by adding either cement or bitumen to the salvaged material. Cement is added at the rate of about 7 per cent, typically by spreading on the surface of the pavement before the stabilising machine is used. Water is introduced at the appropriate rate from nozzles in the mixing chamber, and the fully mixed material is trimmed and graded as necessary before compaction by roller. A curing membrane, typically of bituminous emulsion, is applied and, after curing, a bituminous surface is added. For lightly trafficked roads, satisfactory results have been obtained with the use of fibre reinforced surface dressing (see Chapter 6) on a roadbase of material stabilised in this way. However, the necessary delay before trafficking and the need (in most circumstances) for subsequent treatment are often disadvantageous; the spreading of cement on the original surface can also be awkward, as on windy days in urban roads.

Using similar plant, the stabilising role of the cement may be taken by bitumen or bitumen emulsion introduced to the pulverised material in the mixing chamber. Emulsion is an option which uses a standard recycling machine and which employs commonplace technology to produce stabilised layers up to 150 mm thick; at greater thicknesses the emulsion does not satisfactorily break down to the full depth of the layer. Cutback bitumen is not satisfactory in this application for a number of reasons, perhaps the greatest of which is that the recovered material cannot reliably be coated with such a binder in these circumstances, but the use of a foamed bitumen has been found to overcome these difficulties satisfactorily. Stabilisers fitted with suitable means of injecting the required 2 per cent of water into the supply of pen grade bitumen pumped at mixing temperature from a road tanker are in use in various countries worldwide and are achieving satisfactory results. After processing, the mixed material settles as the foam collapses and may be graded and compacted to achieve the required profile and density. As with the cement stabilised material the surface should be sealed against the ingress of moisture and the abrasive effects of traffic by a suitable surface dressing or wearing course.

Where a stabiliser is not available, or where the scale of the project does not warrant the use of such a specialised item of plant, the retread process may be used. This consists of the repeated scarification of the top 75 mm of the road, until the material has been broken down to approximately the original aggregate grading. The work is then sprayed with bitumen emulsion and harrowed, this latter part of the process being repeated until a sufficient binder content has built up through the whole of the material after which it is graded, rolled and sealed with surface dressing. This method is suitable for use on lightly trafficked roads where the pavement defects are not deep-seated. It is economically viable where the area to be treated exceeds about 5,000 square metres.

Yet another alternative, and one which has been used on major roads in the UK, is the surface regeneration process – also known as 'repave'. Here the upper 20–30 mm of the pavement is heated, planed, relaid and reshaped to be followed immediately with a thin layer of new asphalt or bitumen macadam – perhaps 20–25 mm thick. The effect of laying this on the hot restructured material is that the new surface is securely bonded to the surface to form a new composite wearing course about 50 mm thick, with consequent cost savings of the order of 25 per cent for jobs in excess of about 10,000 square metres, when compared with the cost of simple resurfacing.

9.5.3.2 *Off site stabilisation*

When bituminous materials are removed from a road, they have traditionally either been disposed of, or used as fill material, or for cappings or similar purposes. So-called rejuvenation processes take these salvaged bituminous materials – usually asphalt because of its higher original binder content – and mix them with an appropriate binder such as a 200 pen bitumen to produce a material suitable for use in the pavement. The product may be a low grade base course macadam material suitable for use on lightly trafficked roads, or, where the mixing takes place at high temperature, a material suitable for use as a surfacing. Because of the need to transport material from site to works to site the cost savings here are slightly less than may be expected for the mix in place methods, but there are nevertheless substantial savings to be made in comparison with the use of equivalent materials obtained from conventional sources.

9.6 Repair techniques: rigid pavements

The repair methods which are available for rigid pavements are best categorised by the nature of the defect which they seek to treat. These defects are of three general types; those found at joints; at the surface of the slab; and those of immediate structural significance.

9.6.1 Defects at joints

It will be recalled from Chapter 5 that joints are provided in many concrete pavements to permit the release of thermal and shrinkage stresses, and that these joints generally include a vertical discontinuity through the slab and a tenacious flexible material at the top of the joint whose function is to exclude water and surface dirt while at the same time allowing the joint to retain its flexibility. There are in addition tie bars or dowel bars whose function is to facilitate the transfer of stresses across the joint. This system can be disrupted in either of two general ways.

The sealant material may fail. Flexibility is lost with age and the result can be cracking within the sealant, or a loss of bond between the sealant and one or both sides of the joint. Water penetration can result in corrosion of the steel tie or dowel bars; softening of the subgrade leading to contamination of the subbase, mud pumping and cavitation beneath the slab; and in particularly adverse situations to frost damage at the joints. Dirt penetration will lead to a reduction in the effectiveness of the joint – this will not prevent the joint from opening up as the adjacent slabs contract for any reason, but when conditions change and the tendency is for the slabs to expand a joint which contains more than a little dirt may be unable to contract satisfactorily and the slabs may be damaged. The only treatment which will properly resolve this defect is to remove the old sealant, clean the sealing crack with compressed air, and provide a new seal.

The slab may fail at its edge adjacent to the joint. Modes of failure include arris spalling, shallow spalling, and deep spalling and are illustrated in Fig. 9.10. Of these, arris spalling and shallow spalling are generally caused either by the entry of solid material into the joint as described above, or by defective workmanship at the time of construction. All examples of such failure should be treated whatever their cause since a common effect of these defects is to hasten the failure of the sealant. Treatment should consist of either widening the sealing crack up to about 30–40 mm to include the failed area, and resealing; or cutting out all unsound material and forming a chase of roughly rectangular cross-section alongside the joint and, after taking measures to allow for reforming the joint seal, placing, compacting and curing a suitable repair material.

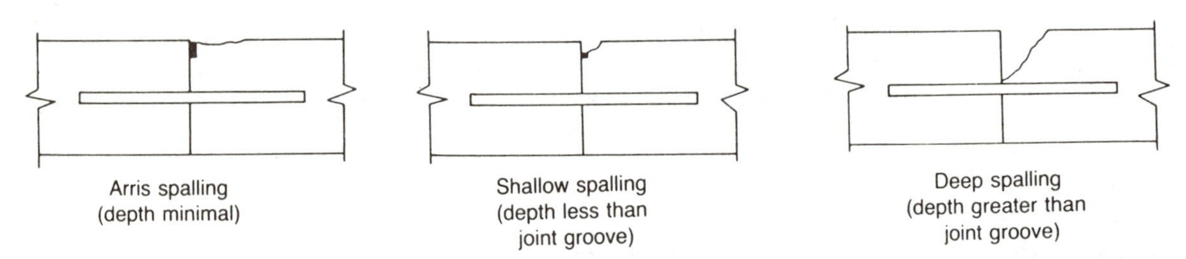

Figure 9.10 Spalling at joints in rigid pavements

Such a repair material is often pavement quality concrete, modified in that the maximum aggregate size is 10 mm. Surfaces should be primed with cement grout immediately before application of the fresh concrete.

Deep spalling penetrates at least to half the depth of the slab and is caused by restrained forces at joints, the restraint provided by solids in the joint or by defective dowel bars which do not allow free relative movement. Very local treatment such as that recommended above for shallow spalling will not be effective in the long term; a more satisfactory repair is achieved by cutting out the full width of the affected slab and making a full-depth repair. Such a repair is illustrated during construction in Fig. 9.11. To avoid a repetition of the failure, care must be taken when drilling for the dowel bars to ensure that the holes are correctly aligned; horizontal drill stands are available for this purpose.

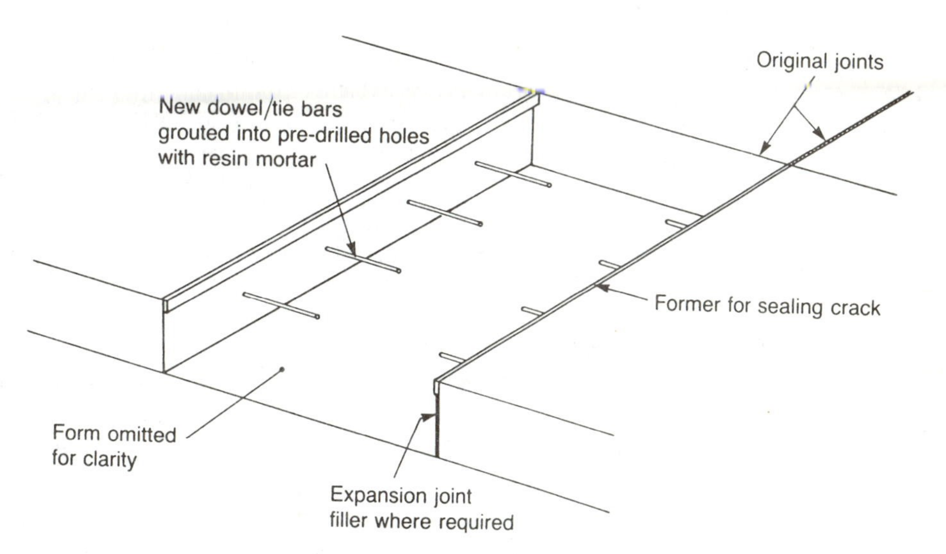

Figure 9.11 Rigid pavement: full depth repair awaiting concreting. Note: for large repairs some reinforcement may be necessary

9.6.2 Surface defects

The most common surface defect of a rigid pavement is the loss of surface texture which in turn leads to poor skidding resistance. The general requirements of a road surface necessary to provide resistance to skidding have been described in terms of the macrotexture (to provide a dry interface between wheel and road) and the microtexture (to facilitate a good grip between wheel and exposed aggregate).

Where the defect is found to be in the loss of surface texture in the concrete slab (poor macrotexture) then this may be remedied by cutting transverse grooves in the pavement surface. These grooves are saw cut and will provide a durable finish to the concrete but the method suffers from the disadvantage that excessive tyre noise can result. To avoid this while ensuring that good drainage is provided the practice is to cut narrow grooves – about 3 mm wide – which are about 4 mm deep, and to arrange the distribution of these along the road in a random pattern, distances of between 30 and 55 mm being left between grooves. The DoT Specification for Highway Works suggests a suitable arrangement of groove spacings.

Where the fault lies in the microtexture of the aggregate exposed at the surface this technique will be of little use. Options which should be considered include surface dressing or mechanical roughening of the surface. Roughening by scabbling or shot blasting will expose more of the aggregate which, if faulty, will fare no better than before. Surface dressing may not be satisfactory at all sites – for example, where there are heavy turning novements – but away from these can be very effective, having performed well on motorways and other well-used pavements.

9.6.3 Structural defects

There are two very general manifestations of structural defects in a rigid pavement – cracks and slab movement.

Cracks are classified by size and by orientation. Table 9.7[6] indicates the relative significance and appropriate treatment for various categories of crack.

Additionally, cracks may form within individual bays – either across corners or between edges and manhole or gulley recesses. These are caused either by defective joints, poorly proportioned slabs, or subbase defects in the former

Table 9.7 Cracks in rigid pavements

	Orientation	
Crack type	*Transverse*	*Longitudinal*
Narrow cracks 0–0.5 mm	Commonplace – not considered to be structurally significant. Narrow crack retains high degree of aggregate interlock and load transfer and allows only a little water penetration. Very low maintenance priority unless in plain concrete	Not commonplace – generally indicative of incipient failure Immediate action necessary only in plain slabs. Treat these by stitching as Fig. 9.12
Medium cracks 0.5–1.5 mm	Significant weakening of the slab, and serious water penetration.	
	Treat by forming a groove at the surface and sealing the crack as at a joint	Treat by stitching as Fig. 9.12
Wide cracks 1.5 mm+	Complete loss of interlock and free entry of water	
	Treat by full depth repair as Fig. 9.11	Treat by full depth repair or by replacing the defective bay
Causes	Bay too long Poor workmanship	Bay too wide Poor workmanship Settlement

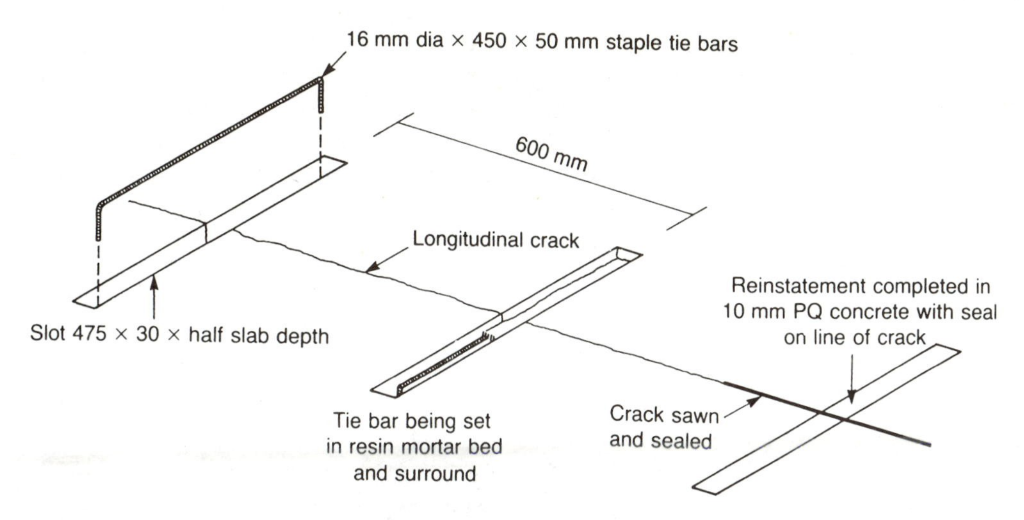

Figure 9.12 Stitching of longitudinal crack in a rigid pavement

case; and by poor design of the relative positions of ironwork recesses in the latter. These should be treated in a similar way to that recommended in Table 9.7 for transverse cracks, the treatment required depending on the width of the crack, as before.

Slab movement is a result of uneven support conditions beneath the slab. These can arise from settlement in the subbase or other elements of the pavement foundation. Equally commonly the cause is found to be cavitation, generally caused by the movement of water after the slab has been constructed. There are two possible sources of moving water. Firstly, penetration through the pavement via defective joints, cracks and so on, causing saturation of the unbound elements in the pavement foundation, which, when the surface is trafficked, forces a slurry up through the pavement, possibly to the surface, leading to a significant long-term loss of material from the foundation. Secondly, there may be running water present below the slab which simply washes away fine particles from the foundation. The first of these is known as 'pumping' and is often detected by the tell-tale presence of muddy water on the surface at a defective joint. The latter is much less easy to detect. If the source of running water is a leaking water main then it is possible that the water authority may become aware of the loss, but water from leaking sewers or from natural underground watercourses may remain undetected for many years before being discovered by chance during routine excavations in the highway for other purposes, or by the failure of the pavement.

Slab movement may be remedied either by replacing the slab, or by filling the voids beneath it which allow the movement. Slab replacement allows the condition of the subbase to be investigated and also permits the replacement of dowel bars which may be defective but is nevertheless often not an attractive option because of the cost involved and the need for lengthy lane closures. Alternatively, the voids which may exist beneath a slab can be filled by injecting material into them.

Pressure grouting has been in use for many years and consists of pumping cementitious or resinous grout through predrilled holes in the slab, the intention being to fill all the voids with the new material which will then set and provide proper support to the slab. This is a specialist process. Possible disadvantages are a lack of control of the area treated, particularly where stiff grouts are used, and the possibility of the grout entering adjacent pipes and service ducts – not all the stories told of telephone cables grouted into their ducts and grout overflowing from downstairs toilets are apocryphal.

Vacuum grouting is a more recent development which permits the admission of a resinous grout below the slab in a more controlled manner. The process is illustrated in Fig. 9.13.

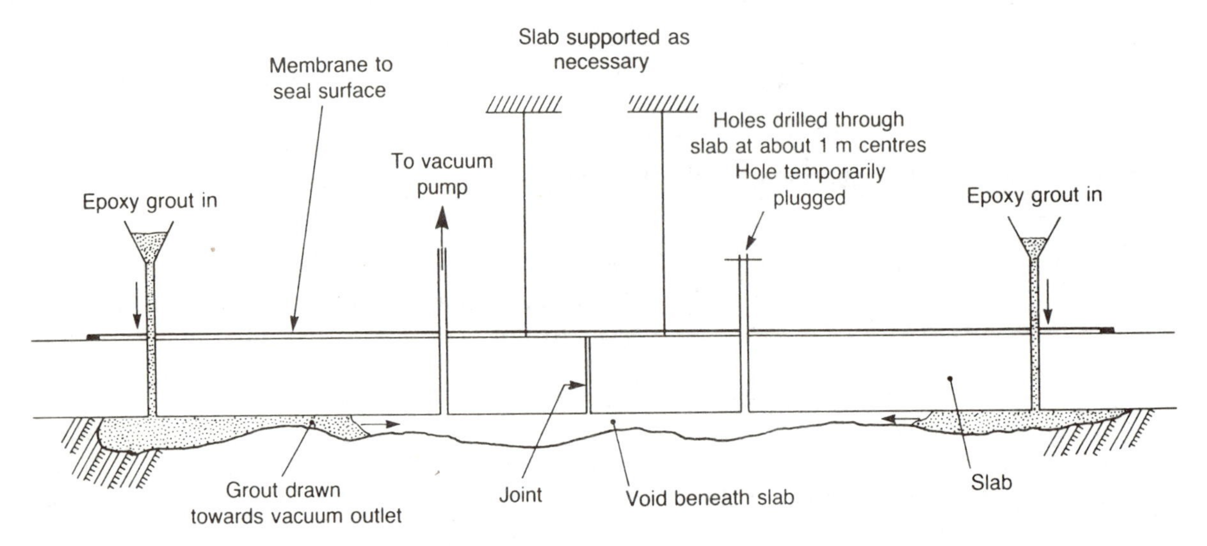

Figure 9.13 Vacuum grouting

9.6.4 Defects in continuously reinforced concrete slabs

Continuously reinforced concrete (CRC) pavements differ from those discussed so far in this chapter in that the pavement is subject to an almost uniform level of relatively high stress, and in that, as there are no joints, there are no points at which repair work can easily be terminated.

Defects may arise in CRC pavements as results of inadequate lapping of longitudinal steel, or of poorly detailed end-of-day joints, or of localised subbase failure. Because of the relatively high stresses which obtain in the slab, the effect of any transverse failure is likely to be correspondingly more severe than would be the case in a jointed concrete pavement. Once failure has occurred, it is also correspondingly more difficult to make satisfactory repairs. Structural defects can only be properly remedied by full depth repairs, a suitable detail for which is shown in Fig. 9.14. The timing of repairs is significant, in that the stresses in the slab vary with its temperature. In the summer compressive forces will be greatest and for this reason the spring or autumn are often favoured; concreting is often difficult in the winter. However, where there is a substantial loss of tensile strength at a transverse crack – for

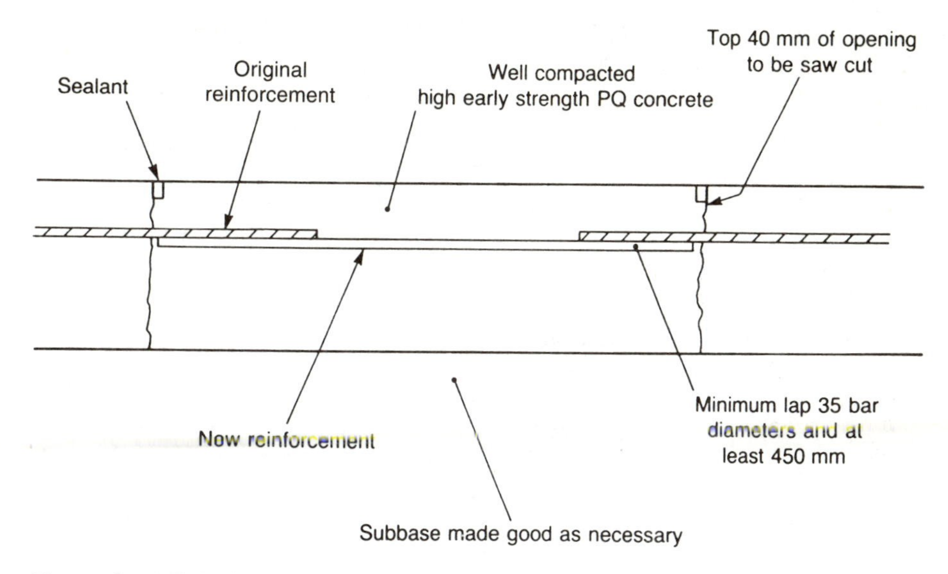

Figure 9.14 Full depth repair in CRCP

example, due to reinforcement failure – a winter repair may be the best option since this will minimise the subsequent tensile stresses in the slab and reduce the tendency for similar failures to happen in the locality.

9.6.5 Overlays on rigid pavements

In cases where the surface of a rigid pavement is showing widespread signs of deterioration, and there are also extensive joint defects, an option which should be considered is that of providing an overlay. Such an overlay may be in pavement quality concrete, or in a bituminous material. Either will also make a contribution to the strength of the pavement.

The provision of an overlay in PQ concrete is generally known as overslabbing. Concrete overlays can, if suitably reinforced, effectively distribute traffic loads to a sufficient extent to minimise the need to stabilise the old pavement before overslabbing; nevertheless, any vertical movement which is apparent in the original surface should be stabilised first. The interface between the old and the new concrete should be partially bonded – new concrete cast direct on the old – where the original pavement is in a fair condition but when cracks are extensive the risk of reflection cracks forming on the new work becomes unacceptable with this arrangement (see Fig. 9.15(a)) and so in such cases the overlay should be debonded by suitable pretreatment of the original concrete. Similarly, joints should be provided in the new work to coincide with those in the original so that the overlay should not be overstressed by horizontal movement below.

With these provisos, the design and construction of the overlay should proceed in a similar way to that previously described for new work.

Bituminous overlays have been used with mixed success on concrete pavements. The common problem is that movement at the joints in the rigid

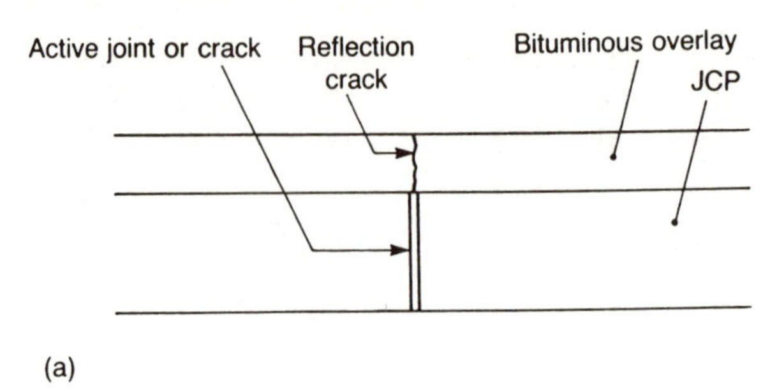

(a)

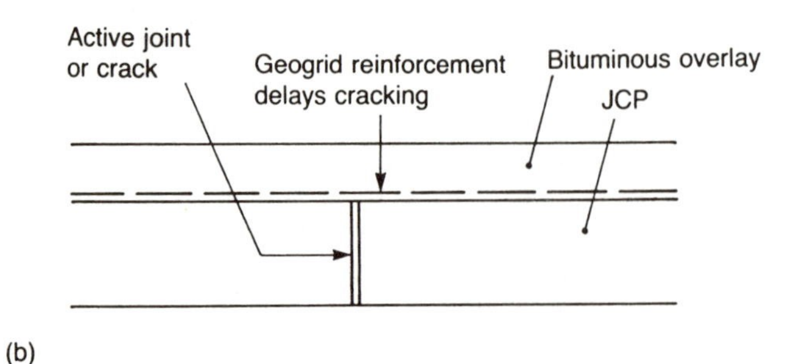

(b)

Figure 9.15 Bituminous overlays on jointed concrete pavements. (a) Unreinforced overlay. (b) Reinforced overlay

pavement reflects through the overlay, in a similar way to that shown in Fig. 9.15(a). The traditional response to this has been to provide thick overlays, or to seal the cracks in the overlay with hot poured sealants as the cracks appear. Neither of these is wholly satisfactory, and there has often been a reluctance to use bituminous overlays for this reason. There are two ways of addressing the problem.

Reinforced asphalt is a relative innovation in the UK, although much more widely used in the USA. The method involves the inclusion of a geogrid in the lower layer of the overlay, as in Fig. 9.15(b), whose function is to provide the overlay with sufficient elasticity to recover from the concentrated stresses at joint; cracks form, as they would otherwise, but the rate of propagation of the cracks upwards through the overlay is severely restricted to the point at which the composite material becomes an attractive design option. Reinforced asphalt is discussed in Chapter 4; the design of the layer may be deduced from FWD tests and the need to provide a minimum thickness of overlay to maintain its own structural integrity – as is equally the case in the design of overlays to flexible pavements.

As an alternative to the modification of the overlay properties, one may seek to modify the properties of the rigid pavement and a successful way of achieving this can be to render the concrete slab flexible by crazing it. One proprietary way of achieving this is by the action of a powerful hydraulic ram, mounted vertically on a substantial vehicle and acting so that the ram applies repeated heavy blows to the original surface, but does not greatly penetrate. Punching the slab in this way at points about 1 metre apart longitudinally and transversely – the exact spacing depending on the slab depth – will result in the crazing of the slab to a sufficient degree to modify its behaviour such that it acts as a flexible element in the pavement, on which a roadbase and surfacing may be laid. The method requires that underground services should not be close to the surface. On lightly trafficked roads a cost saving is offered since removal of the slab is not necessary, and a reduced thickness of blacktop, compared with that which would be necessary for full reconstruction, is required.

Revision questions

1 In the field of highway maintenance, a major technical and managerial problem is that of deciding where maintenance resources are best applied. Describe
(**a**) the way in which the maintenance needs of a flexible road vary throughout its life, and the effects of this on maintenance management
(**b**) two methods of assessing the relative maintenance needs of various individual lengths of a large network of public highway of flexible construction.

2 Outline the way in which the deflectograph can be used to estimate the remaining probable life of a pavement. To what type of pavement is this technique inapplicable? How would you assess maintenance needs in such a case?

3 Describe, in outline, four general techniques used in the structural maintenance of a bituminous pavement. In what circumstances are each of these appropriate?

4 Outline the techniques used in the routine structural maintenance of rigid pavements. Why are these techniques sometimes necessary?

References and further reading

1 Ministry of Transport – Report of the Committee on Highway Maintenance, HMSO, 1970.
2 'Deflection measurement of flexible pavements – operational practice for the deflection beam and the deflectograph', Department of Transport Departmental Advice Note HA/24/83, Department of Transport, 1983.
3 Kennedy, C. K. and Lister, N. W., 'Prediction of pavement performance and design of overlays', Transport Research Laboratory, Laboratory Report 833, 1978.
4 'Benefits of non destructive testing of pavements using dynamic procedures', Halcrow Pavements Group, unpublished report, 1986.
5 Road Note 39, 'Recommendations for road surface dressing', 2nd edition, 1981, DoE/TRRL, HMSO, table 3.
6 Mildenhall, H. S. and Northcott, G. D. S., 'A manual for the repair and maintenance of concrete roads', HMSO, 1986.

Index